The Power of Words

About the Author

Mariano Sigman is an international leading figure in cognitive neuroscience. He was one of the directors of the Human Brain Project, the world's largest effort to understand and emulate the human brain. Sigman has worked with magicians, chefs, chess players, musicians and fine artists to bring neuroscience knowledge to bear on different aspects of human culture. He is the author of *The Secret Life of the Mind*, an international bestseller. He has given several TED talks, which have received millions of views online.

The Power of Words

How to Speak, Listen and Think Better

MARIANO SIGMAN

Translated from the Spanish by Mara Faye Lethem

First published 2023 by Penguin Random House

First published in the UK 2024 by Macmillan Business
an imprint of Pan Macmillan
The Smithson, 6 Briset Street, London EC1M 5NR
EU representative: Macmillan Publishers Ireland Ltd, 1st Floor,
The Liffey Trust Centre, 117–126 Sheriff Street Upper,
Dublin 1, D01 YC43
Associated companies throughout the world
www.panmacmillan.com

ISBN 978-1-0350-4156-5 HB
ISBN 978-1-0350-4157-2 TPB

1 3 5 7 9 8 6 4 2

A CIP catalogue record for this book is available from the British Library.

Illustrations by Javier Royo

Typeset in Warnock Pro by Palimpsest Book Production Ltd, Falkirk, Stirlingshire
Printed and bound by CPI Group (UK) Ltd, Croydon, CR0 4YY

To Fran

Contents

Anyone can have a bad day 1

1. **THE STORIES WE TELL OURSELVES 7**
   How to improve our reasoning
2. **THE ART OF CONVERSATION 37**
   How to make better decisions
3. **OUR LIFE STORIES 73**
   How to edit our memory and discover who we are
4. **THE ATOMS OF THE MIND 127**
   How to clarify what we think and feel
5. **GOVERNING OUR EMOTIONS 171**
   How to take control of our emotional lives
6. **LEARNING TO TALK TO OURSELVES 223**
   How to be kinder to the people we love most

Epilogue: Feynman's Mirror 260
Acknowledgments 263
Bibliography 267
Index 283

Anyone can have a bad day

"It's impossible. You'll be competing against two hundred older boys," my parents told my brother when he said he might be able to place among the top runners in the school race. The next day, Leandro came home with a wide smile and a medal around his neck. A year later it was my turn to race and I expressed the same optimism. After what had happened with my brother, my parents encouraged me without reservation and showed up with a big analog camera to record my feat.

We were a huge group of children in athletic wear at the starting line of a muddy track covered in deep ditches. As soon as the race started it became abundantly clear to me that I was not going to win a medal that day. I was being passed on both sides, at top speed. I was already straggling, going uphill on my way into the forest, when I started to feel dizzy. My legs were weak, my stomach was churning, and, a few seconds later, I was on my knees under a tree, vomiting.

When I mustered up enough energy to stand and walk to the finish line, in last place, I told myself: "I'm just no good at sports." In those days I was a whiz with numbers; my teachers kept advancing me into higher grades for math class to test the limits of my ability for mental calculation. That was where I belonged. I was good at thinking, bad at running: my body was weak, my skin thin, and I didn't have the strength or the mettle for racing.

I settled into that place and developed there over forty years. Until one day, after jogging a couple of kilometers, I felt a pain in my chest. A few hours later I was in the cardiology ward, my body covered in cables. The nurse explained that they'd found several obstructions in my coronary arteries and that they were going to send a stent from my groin into my heart to unblock them. I was shivering with cold as I compulsively repeated that everything was going to be fine. And it was: my arteries were less blocked than they had initially told me.

Back in Madrid, where I had moved not long before, I bought myself a bike. I went out one winter day, wearing long pants and a woolen coat, and rode the most decisive fifteen kilometers of my life. I was pedaling comfortably and had the feeling that I was traveling through nature at the perfect speed. Those fifteen kilometers turned into thirty, then seventy, a hundred, two hundred. One day I was invited to a dinner party three hundred and fifty kilometers from my house and I biked there, as if it were the most natural thing in the world. At some point on that journey, over which I saw dawn break and I crossed forests and mountains, riding alone against the wind, I remembered the character played by Sean Penn in *It's All About Love,* who takes so many pills to get over his fear of planes that he experiences the opposite: he never stops flying, never puts his feet back down on the ground. That's how I felt on my bike.

Some months after beginning this adventure, I went to Morcuera, a mountain with a very steep slope of about nine kilometers. It took me nearly two hours to reach the summit. I came back several times and Morcuera became my benchmark, as it is for many Spanish cyclists. Every time I biked it, I went faster: reaching the top in ninety minutes and then seventy. Fifty minutes, forty-five, forty-two, forty and, finally, thirty-eight. And while that time was much better than I'd ever imagined, I set myself a new goal: to reach the peak in under thirty-five minutes.

I trained hard. I planned for a sunny day that wasn't too hot or too windy. I went by Ángel's bike shop and, while fine-tuning each gear on my bike, he told me he'd ridden up that mountain so fast he hadn't even noticed that there was a lake near the halfway point.

When I arrived at the foot of the mountain, I started pedaling furiously. I was already short of breath and fighting against the sweat irritating my eyes when I looked to the left and, there in the middle of the valley, I saw the lake. I thought of Ángel and imagined the many others who had biked past that same place, their legs burning, as they tried to find their own limits. I wiped my eyes and continued pedaling with all my strength, hearing only the sound of the bike chain until the forest opened up before me and I felt the wind on my face. I was on the last stretch, with only about three hundred more uphill meters to go. I stood up on my bicycle and fixed my gaze on the front wheel, which I moved from side to side with the entire weight of my body. Soon after that, the pedaling finally got easier. I was on the flat stretch. Only then did I lift my head and see the brown sign on two gray stakes that read: "Puerto de La Morcuera: 1796 m." Below me was a narrow, poorly paved road that extended along the plain before disappearing on the other side of the mountain. I saw the dark ground with a few patches of muddy snow, and a couple eating their breakfast at an aluminum table.

I dropped the bicycle and then my body to the ground. I rested for a few seconds, slowly came back to life, and looked at my watch: 32.43. I had crushed my previous time. The sound of those numbers—"thirty-two, forty-three; thirty-two, forty-three; thirty-two, forty-three"—made a perfect chant. I repeated it just as Antoine Doinel had repeated his name in front of the mirror to feel the life in his body.

I couldn't catch my breath. I was exhausted, dizzy, nauseous, about to vomit. After thirty minutes at a heart rate of a hundred and eighty, my body was reacting just as it had when I was eight years old, when in the middle of the race I'd collapsed by a tree. And I remembered what I had told myself then: "I'm just no good at sports."

It took me forty years and thirty-two minutes to understand how wrong I'd been. It wasn't that I didn't have the mettle as a boy. What I lacked was the proper physical condition for racing, either because of my natural predisposition or because I hadn't trained enough. Given those conditions, I had reached my limit. Perhaps, as I should have realized, I'd even surpassed it.

That thirty-two minute and forty-three second ascent of Morcuera retrospectively changed my childhood. I hugged the boy I once was. Tenderly, affectionately, and with a big smile, I apologized for not having honored the effort that he'd made, for not having understood it. It took me all that time to reinterpret an episode that had been the seed of a stigma that I myself had created: "I'm just no good at sports." If I had chosen another phrase, something like: "You just had a bad day, you gave it your all and you can improve," I could have created a very different story.

I'm writing this book because I believe there is no better use of our time and energy than discovering how to change the course of our personal evolution: what we do and don't do, what we feel, and who we are. This project began as a quest to learn more and ended up becoming an introspective voyage, researching those parts of my life where I was most stuck. My fervent hope is that it will also be useful to you. Backed by facts and science, this book will help us create better versions of ourselves.

Our minds are much more malleable than we think. In fact, we maintain—throughout our entire lives—the same ability to learn that we had as children, as surprising as that may sound. What we do lose over time is the motivation to learn, and that leads us to create beliefs about what we can no longer be: some of us are convinced we're no good at math, others that we weren't born to be musicians, or that we can't control our temper, or overcome our fears. Debunking those beliefs is the first step to improving anything, at any moment in our lives.

This is the *good news*: we can change our mental and emotional lives, even the most ingrained aspects of them. The *bad news* is that this transformation doesn't come about just because we intend for it to happen. We must learn how to make good decisions in areas where we've grown used to solving things on autopilot. Just as we come to lightning-fast conclusions of whether a person seems trustworthy, intelligent, or fun, we make judgments about ourselves with equal haste and inaccuracy. That is the habit we must change.

Luckily, the *bad news isn't that bad*. We have a simple and powerful tool: good conversations. This isn't a new idea: almost all Greek philosophy was built through the exchange of ideas in symposia, on walks, and at dinner parties. The great French philosopher Michel de Montaigne put this idea into practice: in a period of brutal conflicts and massacres, he warded off attacks by offering feasts and conversation to those who came at him with sabers.

Today conversation is more alive than ever, in all sorts of media and formats. We can even speak with people in the most remote corners of the world. Yet at the same time it seems to have lost its power; we look down our noses at it, skeptical of its ability to help us improve our thinking. This is unfounded; *good* conversation is the most extraordinary factory of ideas we have in our reach, the most powerful tool we have to transform ourselves, to live better lives.

In recent years, the science of conversation has flourished, and its conclusions should fill us with optimism. It has showed us that through dialogue we substantially improve our decision-making and reasoning and, in general, clarify our ideas and feelings. The reason is simple: exchanging ideas in conversation reveals mental processes that would otherwise go unnoticed. It serves as a *control tower* for detecting errors and glimpsing possible alternatives.

Words have the power to change not only our reasoning but also our beliefs, memory, ideas, and emotions.

CHAPTER 1

The Stories We Tell Ourselves

How to improve our reasoning

REALITY
OUR PERSPECTIVE IS VERY LIMITED
I'M SAD
FALLIBILITY
REFLEXIVITY
CONSOLIDATES AND EXPANDS
WE ARE AMPHIBIOUS CREATURES
PSYCHOLOGICAL BUBBLE
NEWS
VR
DREAMS
FICTION
FAKE NEWS ABOUT OURSELVES
CHILDHOOD GAMES
"WE INHABIT A FICTION"

IT ALLOWS US TO SEE WHAT OTHERS SEE, AND CORRECT OUR FALLIBILITY

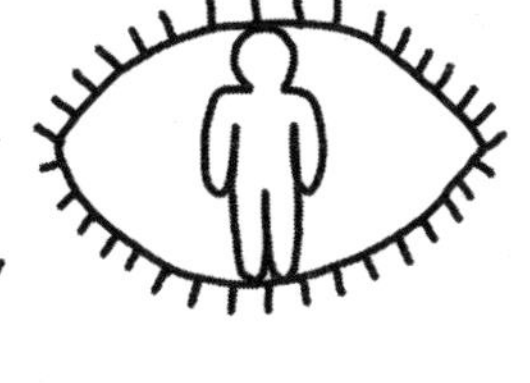

PLUS, YOU LISTEN TO YOURSELF

WHICH IS HOW WE DETECT

AND CORRECT ERROR

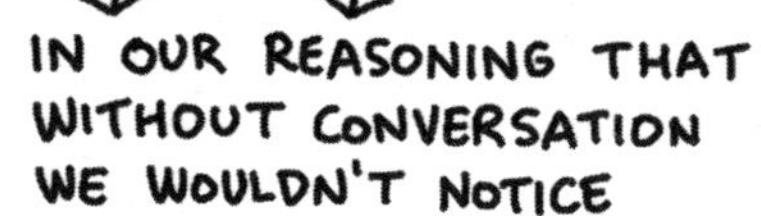

IN OUR REASONING THAT WITHOUT CONVERSATION WE WOULDN'T NOTICE

We respond automatically to complex problems based on the scant information we have at hand. We meet someone and, in less than a second, we form all sorts of beliefs about them. Since we are not even aware of all the considerations and arguments we aren't taking into account to reach these conclusions, we trust them fully even though we may be mistaken.

This bias makes language a double-edged sword. On one hand, its capacity to combine words endows it with potentially unlimited precision. In practice, however, we never use this resource. Instead we end up communicating in very rudimentary ways. For example, we often describe nuanced emotions in few words, and this limits our ability to recognize and differentiate a vast continuum of desires and stressors. When a complex feeling is summed up in a phrase such as "I'm sad," or "I like Juan," the sentence itself becomes a filter through which we perceive reality. This is language's reflective property: the capacity of a statement to modify what it describes, above all to the speaker.

Our partial and distorted view sometimes makes it difficult to distinguish between what is true and what is false, something that is now common currency as fake news. But this short-sightedness is not unique to distant worlds. It is rather an idiosyncratic feature of our cognition: lies merge with the truth, and in that mixture—the fake news about ourselves—we create our personas over time.

Given how these biases in our cognition lead us to all kinds of mistakes, I propose a solution: learning to converse. This ancestral tool, so simple yet so powerful, reveals our errors in reasoning that often go unnoticed. Dialogue allows us to resolve them, thus substantially improving our way of thinking.

On April 15, 2013, shortly before three o'clock in the afternoon, two bombs exploded amid a celebratory crowd near the finish line of the Boston Marathon. Those responsible for the attack made an escape straight out of the movies. It included kidnapping a driver, throwing homemade explosives, killing an MIT campus policeman, and shoot-outs in residential areas of the city. The Boston marathon bombing was one of the first news events to be transmitted in real time on social media, and Soroush Vosoughi was one of its first spectators. From his desk at MIT, he saw the drama unfold simultaneously outside his window and on Twitter, and he understood something that would soon be clear to everyone: it was almost impossible to separate truth from fake news. The virus of language found its perfect breeding ground on the social networks.

The Power of Words

Vosoughi ran through the hallway that led to his thesis supervisor's office and told him straight away that he wanted to change the subject of his doctorate. Then he began developing a tool to detect the veracity of the rumors circulating on Twitter. In an unprecedented computational effort, in the era just before big data, he analyzed millions and millions of tweets with opinions and facts about sports, politics, celebrations, love, envy, hatred . . . His objective was to develop an algorithm capable of separating, in this seemingly infinite database, the true statements from the false ones. Could it be that fake posts are usually shorter? Do they have more exclamation points? Are certain words more likely to be part of a lie than the truth? Is it the message or the messenger that confers credibility?

A few years later, these questions (and many of their answers) have become common currency. However, in those days, the discovery made by Vosoughi and his team was highly surprising. The best indicator of whether a tweet is truth or false is not in what it says or how it's written, or who wrote it, but in how we react as readers.

Lies are easily recognizable because they spread like wildfire. Vosoughi noticed this across many fields: politics, ideology, sports, gossip. Fake news spreads "farther, broader, and faster"[1] than the truth.

We are more likely to spread fake news because it is not bound by the circumstantial limits of reality. And that freedom allows for an exaggeration of certain aspects, including the emotional, which is highly attractive to the brain. William Brady, a researcher at New York University, discovered that each emotional word made a tweet travel at an impressive 20 percent faster rate.

Why do we believe lies?

These strategies are not unique to fake news. They are common to all the fictions that surround us. The increased heart rate of a person walking on the edge of a cliff in a virtual reality game is indistinguishable from that experienced in real life. Our bodies confuse truth and falsehood, reality and fiction. The worlds on either side of our devices coexist in a very particular way. Sometimes we are so immersed in the virtual experience that we almost forget it isn't real. But if anyone ever asks us where we are, we don't hesitate to answer that this illusive

1 It almost sounds like a Vin Diesel movie.

universe is nothing more than a game. We are amphibious creatures: we enter and exit fiction like a frog hops in and out of water, without effort and sometimes without even realizing we've made the shift. Throughout the evolutionary process, some amphibious species lost that ability over time and became inhabitants of a single environment. Will the same thing happen to us? Will we lose the amphibious quality that allows us to move between fiction and reality? Will fiction end up becoming the definitive habitat for our species?

In virtual reality, the term *presence* is used to describe those moments when we merge completely with the fiction. What is it that leads us to forget that the virtual world is merely an invention? The answer is not what you might suspect. It has little to do with faithful reproduction of every last detail. Mavi Sanchez Vives discovered in her VR laboratory that what causes presence is not the sensation of "being there" but rather of "doing there." Or, as was proposed, long before all this technology, by Martin Heidegger about the human experience: *dasein* or "there being." This abstract concept appears in our lives, from childhood games to dreams. When a child sits astride a wooden stick, they know it isn't a horse—in fact, if it did turn into a real animal, the child would be very frightened. It is their riding and the enemies they chase or flee that impart presence to that world represented by a mere broomstick.

Fiction does not require the creation of a world similar to the one in which we live. That doesn't matter to anyone. That is not what confers presence. A classic example is seen in cinema, with black-and-white films: no one has ever seen the world in black and white except in photographs or on the silver screen, yet that chromatic improbability not only doesn't preclude presence but sometimes even instills it. In the world of magic it is even more evident. We all know that the silk handkerchief the magician makes disappear hasn't really become a dove, but that doesn't affect our amazement. Presence is often associated with the "willingness to momentarily suspend disbelief" but this idea is not quite accurate because presence is involuntary. It happens despite us, as part of our amphibiousness.

In literature, presence confers internal coherence, thanks to which the narrative flows and readers can immerse themselves without interruptions from the other side of their amphibious world. Jorge Luis Borges summed up this idea better than anyone else:[2] "What does being a writer mean to me? It means simply being true to my imagination. When I write something, I think of it not as being factually true (mere fact is a web of circumstances and accidents), but as being true to something deeper. When I write a story, I write it because somehow I believe in it—not as one believes in mere history, but rather as one believes in a dream or an idea."

The broomstick horse is just one of the many creative inventions of childhood. The line separating play from fiction is a very thin one, and children's daydreams often turn into lies when they transcend the child's personal realm. That is when they are subject to the skeptical gaze of others, and reality begins to demand explanations that require increasingly more acrobatic juggling.

Everyone has their memories. One of mine dates back to 1982, when it was announced that Diego Armando Maradona was joining FC Barcelona. He was going to arrive in the city where I had been living for the past six years. My classmates at school asked me if I

2 As hard as that may be for you to believe!

knew him and, without hesitation, I emphatically said I did. The whole thing wouldn't have gone any further if not for the unfortunate fact that, of all the possible homes he could have chosen, Maradona ended up living just a hundred meters from our school. The pressure from my classmates to introduce them to *my friend* Maradona was so great that I finally gave in. One winter morning at eight o'clock, right before we went on a field trip, we all went over to visit him. My attempts to convince the guards at Maradona's house of our friendship, by shouting that I was Argentine too, didn't work, and the lie crumbled *ipso facto,* without algorithms, without Twitter, without spreading. It had come crashing into reality, and I crashed with it.

Long after that, in Vancouver, I shared the TED stage with Kang Lee, a professor of social psychology at the University of Toronto, who told a story about the lies that children tell. When do they start lying? And, more importantly, why? Kang explained that lying is part of a fundamental cognitive exercise. Lying is practice for understanding others; particularly the difference between what you know and what others know, something that in psychology is known as *theory of the mind.*[3] My friendship with Maradona was an exercise in fiction: a way to create coherent and realistic stories, to erode reality so the narrative was more intriguing. The bubble of fiction grew until it burst there at the door to his house.

Fake news about ourselves

The greatest virtue of words is, at the same time, their greatest stigma. Their fabulous ability to build coherent worlds allows us to express

3 One afternoon, shortly after I'd arrived home, my three-year-old son Noah asked me if Lesly, the person who took care of him while we were working, had left yet. I said no. He asked me again thirty seconds later, insisting twenty seconds after that, again and again until finally, five minutes later, I said yes, Lesly had gone. Then he looked at me with a smile that would have given him away even without all that context and said, "I already took a bath." I was moved by the contradictions in his young mind. He had the sophisticated cleverness to understand that there was only one witness who could contradict his story, yet the naivete to be so obvious when making sure that witness was no longer present.

our fears and desires, but it also imbues our stories with their own momentum. Fake news is more contagious than real news, especially when the story told is about us. "I'm sad," or "I'm happy," or "I'm anxious." Each of these phrases does much more than describe an emotion. They are sentences, like those handed down by a judge. Their interpretations result in actions that influence and condition the very universe they are trying to describe: they can be *fake news about ourselves.*

The famous investor George Soros tested the limits of this idea in one of the most fascinating laboratories of human behavior: the financial market. Soros had studied economics and was a student of Karl Popper, one of the greatest philosophers of science. That academic journey gave him two great principles that were decisive tools for his comprehension of the financial market: *fallibility* and *reflexivity.* Fallibility establishes that people's ideas about "the world" never correspond exactly with the reality. No theory or general opinion is without distortion: they are all necessarily imperfect. And here is where reflexivity comes in: once the theory has been formulated, we act as if it were true and that gives it more weight. It becomes a self-fulfilling prophecy.

In the financial world these two rules come together in a classic example. Investors build the belief that a certain stock is good value. If reality contradicts that belief, they will normally lose money. That is the principle of fallibility; they make a bad bet and they lose. Sometimes, however, that failure doesn't occur because the principle of reflexivity intervenes. The belief shared by all those investors governs their behavior, sweeping away all doubts; they continue investing blindly. As a result, the stock rises and becomes, at least for a while, a good investment. That loop feeds itself and gives rise to financial bubbles, those hotbeds of the market that Soros understood better than anyone. When a bubble grows at a frantic rate, it is not fueled by the universe of companies and their products or by the universe of technology. It is in the reflexive force of investors' beliefs. The price of the stock goes up, reinforcing the investors' enthusiasm, which inflates even more than the price, in a loop that detaches the financial world from the stocks it represents. It seems that this could continue *ad infinitum*. But that is not the case, because, at some point, as with my fib about Maradona, the juggling is eventually not enough and the bubble crashes into reality. And bursts.

Fallibility and reflexivity were studied long ago in psychology and sociology. Soros's contribution was trusting in those principles, highlighting their relevance, and putting them into practice in the world of high finance. Here I borrow them with the same premise: highlighting their relevance and putting them into practice in order to understand the way we think.

We can revisit the sentences that describe our feelings, such as "I'm sad," "I'm happy," or "I'm anxious," in the light of the two principles we've just explained. These *theories* are biased and distorted. The principle of fallibility wouldn't be so problematic if we weren't utterly unaware that our declarations about what we feel are necessarily imperfect, which can lead us to confuse frustration for anger or fear for anxiety. There's no need for much philosophy to explain this. Aventura sang it clearly in her hit "No es amor": "No, oh no, what you feel isn't love; it's obsession. An illusion that makes you do things: that's how the heart works."

It gets worse though. These confusions not only go unnoticed, but are magnified by reflexivity. The sentences we mentioned do much more than describe an emotion: they influence and condition what we feel. They are fake news about ourselves that, once enunciated, have a reflexive power capable of creating a "psychological bubble." Merely thinking that we are angry ends up making us angry: the self-fulfilling prophecy in the universe of the mind.

The phrase "I'm feeling sick today" is a theory that attempts to describe, in a simple way, a whole body of complex data. Like any other conjecture, it is only partially true and has to be fine-tuned and clarified. Perhaps what we are really feeling is sleepy, worn out, or bored. The theory can change, just as physics changed when Einstein proved that Newton's law breaks down when bodies in motion approach the speed of light.

The difference is that one theory applies to things and the other to people. And only in the latter does reflexivity find its breeding ground, establishing a feature distinctive to human sciences as opposed to natural sciences. Newton's law completely changed our understanding of body motion. A revolution that, nonetheless, does not change one iota the way those bodies move. On the other hand, Marxism, liberalism, and theories about inflation or racial supremacy can decisively alter the economic and social universes they purport to describe.

The most convincing way to see the fantastic power of words is, perhaps, when they create that which they describe. Literally. In general, the world takes precedence over the word. When we say, "It's raining," it is the weather that leads to our statement. However, when a judge says, "I sentence him to ten years in prison," the words precede, and generate, a new reality. They don't describe the world, they create it.[4]

---

4 In his book *How to Do Things with Words,* John Austin defined such statements as performative, because, rather than describe actions, they perform them.

Spider-Man's famous maxim, "With great power comes great responsibility," can be perfectly applied to the world of words, because language is binding and has the capacity to construct and transform our mental experience. Words can calm and heal, but they are also capable of creating stigmas and causing illness. "I don't like that," "I'm no good at this," "I can't do it." We blithely broadcast fake news about ourselves—both in our heads, and out loud—not realizing its pivotal role in opening and closing doors for us. Fake news also spreads like wildfire within our own minds. Just as a poor interpretation led me to prematurely decide I was no good at sports, a single phrase can convince someone that they can't paint, learn math, or love. A single phrase can also awaken enthusiasm, banish fears, or convince us that we are capable of seemingly impossible feats.

The myopia of reason

Our tendency to alternate between fiction and reality has its roots in a more basic principle: constantly seeking out explanations for the unknown.

When faced with the series 2, 4, 6, most of us automatically conclude that it should continue with 8, 10, 12, 14 . . . Based on just a few elements, we deduce the rule that seems most simple and obvious. Of course, there are an infinite number of different compatible explanations. For example, the series 2, 4, 6, 12, 14, 28, 30, 60 . . . adding two, then multiplying by two and so on. But the first explanation seems the most evident, and we convince ourselves that that's how it must be. This is not exclusive to number series. Stereotypes, to give a very different example, are rooted in the same principle. The idea that Chinese people are patient, Italians are extroverts, the French have refined palates, and Germans are punctual is a generalization based on individual traits observed in a small sampling of individuals and what we've been told about them.[5] Generalizing and constructing rules on the basis of very little information are the two main causes of social bias and prejudice. They are part of the system

5 Would you celebrate your purchase of a French car with German champagne?

of intuition that leads us to make erroneous decisions we blindly trust. Here we see the darker side of this mechanism that is ubiquitous to human cognition. Identifying rules from scant data is an extraordinary feat. It allows us to immerse ourselves in unknown worlds and very quickly establish principles that help us navigate them. Yet, at the same time, it is a way of tinting reality with fiction, by taking the data and coming up with simple—but not necessarily true—explanations. This trait so particular to human thinking is a result of three factors.

The first is a limitation: except in exceptional circumstances, we only have access to very partial views of the things that surround us. This is true in every aspect of life: the world of objects, of ideas, and of people. When we are forced to choose where to vacation, what to eat, who to vote for, which neighborhood to live in, we usually only consider what we've seen, what we've been told, or what we've extrapolated from a similar experience.

The second is a virtue: our brain is rapid and particularly efficient at coming up with possible rules based on limited data, and that helps us to function without getting lost over and over again in new, unknown worlds. The brain reaches very rapid conclusions that, on average, are of high quality. They are almost always correct. But every so often they lead us to serious failures.

The third is an illusion: not recognizing that, throughout this entire process, our brains steer us to forget that there is an enormous portion of the universe we do not see. Our vision is inevitably partial, but we usually feel and act as if that weren't the case.

Let's illustrate with a game how these principles apply to simple problems. They apply also to other problems that impact every aspect of our lives. We will begin with a logic problem used by Hugo Mercier, a cognitive neuroscientist dedicated to unraveling the enigma of reason. It goes like this:

1. John looks at Mary. Mary looks at Paul.
2. John is married.
3. Paul is single.

The question is: does it follow from these statements that a married person looks at a single person?

There are three possible answers: "yes," "no," or "there is not enough information to know." What is the correct response? It is worth trying to figure that out. Reasoning is a good way to observe our thoughts in a mirror, a good way to discover—in the first person—how we think.

When I was presented with this problem I responded that there is not enough information. In fact, I felt confident and proud of my choice. Understanding that there can be insufficient data for drawing firm conclusions is an essential part of scientific thinking. But, like the vast majority, I was mistaken. The correct answer is yes. One can deduce that a married person looks at a single person. The key is thinking about Mary. While we weren't given any information about her marital status, it is not that unclear: she is either married or single. There is no other option, as Aristotle proposes in his law of the excluded middle.

Let's take a closer look at each case.

If Mary is single, then someone married looks at her: John. If Mary is married, then Mary looks at someone single: Paul. So we know for sure that in this case there is a married person looking at a single person.

This is one of the many examples where our reasoning leads us to rush to a mistaken conclusion. In this case, our attention and thinking are focused on what we haven't been told about Mary, and, once this idea is established, it seems evident that there is not enough information to resolve the problem. We will see how conversations can break down these logical traps and, as a result, allow us to think much more effectively. But first we will review the errors we often make when reasoning in order to convince ourselves that finding a solution is truly necessary.

There are countless problems we could use to illustrate this. If there is a pile of seven apples and you take two, how many apples do you have? What is red and smells just like white paint? If a white horse enters the Black Sea, how will it emerge? If you are in a race and you pass the car in second place, what is your position? The next problem is perhaps a better illustration, since it is itself an illustration. The challenge consists in connecting the nine dots by drawing four straight lines without picking up your pencil.

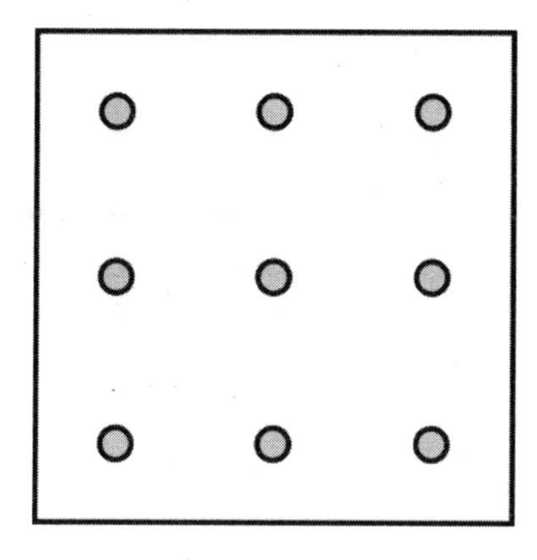

It seems impossible. We come to that conclusion very quickly, after merely a few attempts. However, it can be done. Try to find a solution before checking the answer at the end of the chapter.

The mental trap here is assuming that the lines cannot leave the square. If we realize they can, the problem becomes much easier to solve. Often a single clue can unlock something that, without it, seemed impossible.

Exercises such as these have been the basis of many psychological theories about how the human mind solves problems and makes decisions. The most famous, put forth by Daniel Kahneman, maintains that thought is divided into two systems. The first type oversees almost all our everyday activities. It is quick and automatic and is the one that, based on a small amount of data, rapidly generalizes without giving us any warning. It is also the type of thinking that leads us to errors in the kinds of exercises we listed above. The second type, on the other hand, is slow and requires conscious effort. This is the type of thinking that calculates all the directional options for Mary's gaze and sees past the limits of the dots-in-the-square problem. It never takes charge spontaneously, but it is much more accurate and resistant to cognitive biases. To avoid abstract terminology, I will call the first system *automatic thinking* and the second system *logical thinking.*

Causality and chance

The primary reason we are so often mistaken in our decision-making stems from considering only the evidence we have at hand: the evidence that is *available* to the mind and fuels a much larger set of facts and considerations. This "small error of reasoning" has consequences in practically every aspect of our lives. Let's look at some examples.

The first is a feeling we've all experienced when we talk about someone we rarely see and then the very next day, boom! We run into them in the street. The encounter seems magical. A coincidence bordering on the miraculous. The evidence that is unavailable to us, in this case, is the enormous number of times that this doesn't happen. In other words, when we talk about someone we don't see often and the next day . . . we don't run into them. Since there are so many improbable things, the probability that one of them will happen is actually quite high. But we only *see* what is in front of us, and as such the coincidence seems surprising, almost incredible.

Another example: when we learn about a disease, the probability of being diagnosed with it is magnified in our limited universe of available evidence and, therefore, our fear of it increases. Symptoms that we would have otherwise ignored suddenly seem related to this *new* disease (i.e. newly available to our mind). Seeing only part of the evidence also produces distortions in the story we tell ourselves about our accomplishments and misfortunes, about the inevitable mix of causality and randomness, of winds blowing in and out of our favor, that accompany us throughout our lives.

Helen Pearson, editor for the journal *Nature*, tells of an experiment that sums up this idea in a very clever way. A group of British scientists followed the evolution of almost every child born in England, Scotland, and Wales in one week. They meticulously collected information from every village, neighborhood, and remote area of the islands on the pregnancies and births and all sorts of aspects of the children's first few years of life. They also gathered samples of placentas, locks of hair, nail clippings, teeth, and DNA, and, with the same meticulous scrutiny, looked at the children's lives, families, and social trajectories. The ambitious goal of this project was to understand how this complex mix of biological, cultural, economic, and environmental factors determines one's development. Pearson, with a sense of humor that highlights the extent to which we ignore the role luck plays in the direction our lives take, sums up the main findings of this encyclopedic study: "The first lesson for successful life, everyone, is this: choose your parents very carefully. Don't be born into a poor family or into a struggling family [. . .] because those children tend to follow more difficult paths in their lives." This ambitious experiment revealed that the luck of which family one is born into is the most decisive factor in the geometry of one's destiny.

This may seem to be a very roundabout way of reaching a conclusion so apparently basic. It does, however, contain a paradox: when you ask someone for the reasons behind their accomplishments, the answers almost always refer to hard work, ability, perseverance, risk, or the influence of mentors . . . Rarely does anyone mention the decisive role of luck. In part because that information is not usually available.

This blindness is a recurring cause of disagreements that lead to grudges and disputes. In sports, for example, the fans tend to overestimate the injustices suffered by their team, which makes them

feel they are victims of a conspiracy. Another example, which many readers will recognize: who takes care of the household chores? What percentage does each person do? It turns out that everyone believes they are doing more than they really are. We underestimate the barriers, obstacles, and crosswinds that others face.

To sum up, one of the most frequent errors in our thinking is a result of forgetting that we are always working with a very partial view. This affects how we reason, how we regard our daily lives, political beliefs, and relationships. The problem is further aggravated when the ideas we construct are manifested through language, which gives them a much greater patina of apparent truth than they should have. When the fallibility of our reasoning is exacerbated by the reflexivity of language, it deepens these errors and perpetuates them.

These principles govern all mental experience. Automatic thinking concludes that it is impossible to run four lines through all of the nine dots. And we make the same mistake when believing that there are emotions or ideas that we cannot rise above. This principle is at the heart of every stigma we create. Overcoming these obstacles requires the mediation of logical thought. The question is: how can we summon it? How can we appeal to reason in the impulsive world of emotions? We will find the solution in what Michel de Montaigne, in his most famous essay, calls "the art of conversation."

The Value of Conversation

We have seen that language can devolve to a point of madness in which conversations heighten differences more than they attenuate them. As happens with fake news, which spreads like wildfire, conversations can proliferate confusion rather than reason, they can polarize and stoke hatred, building more walls than bridges. These sorts of exchanges have become so omnipresent that many believe they are the inevitable fate of all conversations, that there are certain topics it is impossible to talk about. This presumption is wrong. When conversations happen in the proper context, with a small group of people listening to each other and expressing their sides, it helps us to think more clearly, to make better decisions, and to be more even-handed, empathetic, and understanding. It's as simple as that: conversations are a wonderful tool, perhaps the most effective one we have for shaping our thinking.

Hugo Mercier brought his logic and reasoning problems into the realm of conversation. After each participant found *their* solution, they gathered in a group to exchange ideas. When reasoning from the distorted lens of automatic thinking, most of them arrived at the group with the wrong idea. Only one of them, or two at most, had reached the correct conclusion. Mercier was intrigued to see who would win that battle. What happens when a minority with the right answer joins a multitude convinced of the opposite?

The experiment is conducted in turns. People converse and then are allowed to change their minds. The model is repeated: they go back into conversation groups, review their opinions, and so on. In some cases, the larger group of mistaken people win out over the minority and make the few who had correctly solved the problem change their minds. This is the greatest risk of conversations: that the social pressure of the majority crushes the good arguments of the minority. In order to understand when that risk is heightened and when it is lowered, let's look at one of the most influential demonstrations of social pressure, conducted by the psychologist Solomon Asch shortly after the end of the Second World War. All he needed were some lines drawn on two poster boards.

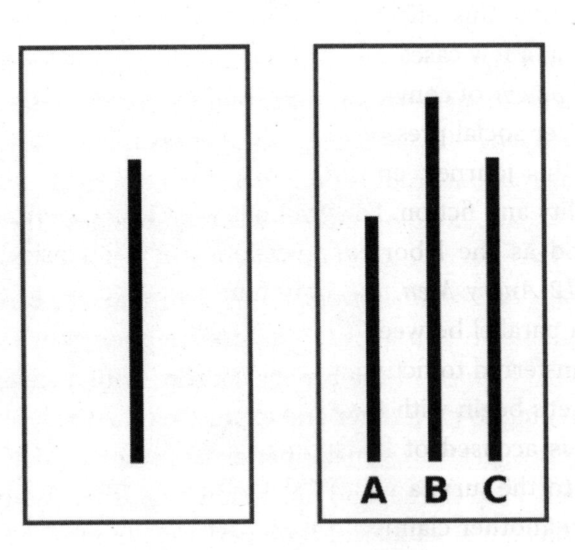

Which of the three bars in the image on the right has the same length as the bar on the left? This problem would have been quite easy to solve were it not for this trap that complicated it: participants were asked to resolve it in groups of eight people, seven of whom were actors who sometimes gave an emphatic, unanimous, and incorrect answer. Would they be able to influence the only participant who was responding truthfully according to what he or she saw?

Asch showed that sometimes people would cave to the majority, proving that social pressure can lead us to defend something totally absurd. As is often the case, the study's results have been distorted; it is important to clarify its true magnitude and avoid exaggeration. Only 5 percent of the people blindly followed the majority in each one of the rounds. And 25 percent—a larger amount but still low—ignored the actors every time. On average, the actors managed to influence the unwitting participant on one out of every three occasions. These results show that the tension between social pressure and reasoning has a large gray area.

I contain multitudes

The risk in conversation comes from our propensity to allow the majority voice to trample all over our ideas. Mercier's experiment shows us that this effect can contaminate the process of group reflection in a few cases. We will now see the flip side, the opposite force: the power of conviction, which in good conversations usually prevails over social pressure.

Along this journey, emulating the way our minds work, I will blend reality and fiction. For example, Mercier's experiment can be understood as the laboratory version of the historical trial that inspired *12 Angry Men*, the only film produced by Henry Fonda. There is a parallel between these two cases: one with the drama of reality transferred to fiction and the other under the microscope of science. Let's begin with the trial.

A boy is accused of killing his father. The judge presents the evidence to the jury: a neighbor says he watched the scene from a window; another claims he heard the boy threatening to kill his father. The son's criminal record doesn't help his case; he'd been arrested for assault and possession of a knife similar to the murder weapon. The judge explains that if they unanimously agree that the evidence is conclusive, the boy will be sentenced to death by electric chair.

The starting point is very similar to Mercier's experiments. Eleven of the jurors are immediately convinced that the boy is guilty. That is the natural conclusion of automatic thinking: they act as if there was no other consideration beyond the evidence they've just heard. The jury sits around a long table. They all agree on the *obvious*: the defendant is guilty. All except one, juror number eight, played by Henry Fonda himself, who wonders if there isn't a margin of reasonable doubt that should be considered. Fonda is alone against all the others. He is a true Hollywood lone rider.

How, when, and why do some people manage to convince a group? Throughout this book these questions are not posed in their classic context: that of leadership and mass phenomena. Above all, I am interested in them because convincing others is not that different from convincing yourself. And the reason is simple: our minds work like a court of opinions. As Walt Whitman put it: "Do I contradict myself? / Very well then I contradict myself, / (I am large, I contain multitudes.)" Hal Pashler and Ed Vul turned this idea into an experiment where they recorded the vast variety of responses we give to a single problem as if they came from different voices and ways of thinking. This is why it is so important to our own thinking that we understand how opinions are formed in a group. *Talking to others is the most natural way to learn to talk to ourselves.*

The power of minorities

The conversation among the jury members is superficial and hasty. They take a preliminary vote with the intention of concluding the case. All of the jurors vote guilty, except for number eight. We can see the parallel between that group and our own voices: on one hand, there is a superficial and impulsive majority that wants to have an open-and-shut case, out of laziness, and, on the other, a more thorough minority that sends out a warning signal: we may be being too rash.

Fonda's character questions the validity of the evidence and, in a master class on logical thinking, enumerates the many errors in reasoning they may have made. Since the rest of the jury is stubborn and impatient, he calls for a new secret ballot from which he will abstain. If all eleven jurors again agree that the boy is guilty, he will accept their decision and the trial will end. The others accept his all-or-nothing proposal. When the votes are revealed they see that one has changed to not guilty. Without this turning point, the real trial would never have been made into a movie. Where is the point of no return? What is the critical mass needed to change a collective belief?

The mathematician Andrea Baronchelli, based in London, answered this question with an online game. Players see a face on their screen that repeats round after round and they give it a name. In each round they are randomly assigned a series of partners who, without speaking, have to agree on the name the face will be given. When each round is over the participants see what their partner has chosen and thus have a very limited sample of what the others are doing. This scant information is enough, after just a few rounds, for the players to come to an agreement and start to give the face they observe the same name. This social convention emerges spontaneously, without the mediation of an institutional mechanism. We see—in the experimental realm—an idea put forth by many philosophers: the meaning of words is constructed through agreements disseminated among peers.

Once they've come to an agreement, the most relevant part of the experiment begins: the players are joined by a small number of participants who've been planted to campaign for another alternative and to try to upend the established consensus. If the group of "confederates" is at least 25 percent of the population, then they are able to convince all the others. Their strength lies in the coherence of their actions, stubbornly maintaining the same conviction amid a more indecisive majority. That is the key: *the power of small groups does not stem from their authority, but from their commitment to the cause.*

The gutters of history

In the trial, the character played by Fonda is initially alone in his ideas. But his passion for reason (it sounds like an oxymoron, but it isn't) turns out to be contagious. He gradually chips away at each argument, finding its inconsistencies, and eventually convinces the jurors one by one. In the plot's hostile asymmetry of one against eleven, he lights the fire with a tiny fuse.

Mercier, in his experiment, discovered a process almost identical to what happened in the cinematic trial. In most groups, the lone rider manages to promote his idea just as Fonda did: by convincing one person, then another, and those in turn convince the others. So, after a few rounds, reason has infected the entire group.

It is time to compare the various results we've seen up to this point. The experiments by Mercier and Baronchelli coincide in the ability of minorities to change group opinion, and they also agree that the minority's power comes from their persistence and conviction, not from a privileged social position.

They diverge substantially, however, regarding the critical mass needed to spark the flame. In Mercier's experiment it only takes one person; in Baronchelli's, a quarter of the total population is required. The explanation for this difference is found in conversation. In the second experiment there is no discussion; in Mercier's, on the other hand, they put forward arguments.

Conversations work in certain arenas, as Montaigne outlined in his *Art of Conversation.* When all of the participants have the time and the right to speak and be heard, the conversation takes on its full force. That is when it becomes an ideal space to assess our reasoning and notice any possible failures: it is the system of logical thought to review the automatic system, i.e. Fonda against the other eleven jurors. As soon as the conversation strays from those conditions, either because there are too many participants or because the desire to listen wanes, we start to see the conclusions found in Asch and Baronchelli's experiments. Those conversations become a social bidding war of intimidation to convince the others through social pressure instead of argumentation. We often see these brutal wars on social networks.

The cases created by Mercier and Baronchelli are extremes. On one hand, pure reason; on the other, pure arbitrariness. In general, discussions take place in mixed environments of reason and convention. This balance of forces changes over time, with argumentative disputes that become entrenched; future generations take up these positions as a new starting point for the conversation.

The Bulgarian philosopher Tzvetan Todorov, in his book about the conquest of the Americas, analyzes the heated discussions in that critical moment in history. In the year 1550 the Valladolid debate took place. This was a formal disputation in which two antithetical views on the humanity of indigenous peoples faced off against each other. The philosopher Juan Ginés de Sepúlveda argued that they were savages and therefore their enslavement was justified. Today that position is untenable, but in that century it held its own against the opposing view, represented by Friar Bartolomé de las Casas, for whom there were no human hierarchies.

History is riddled with discussions whose arguments are incomprehensible in today's terms. It is worth remembering that many of the principles we now take for granted will surely seem absurd in another historical moment: we are the Sepúlveda and Bartolomé de las Casas of the future.

Exercise: Ideas for living better

The label "self-help" gets bad press among many of my fellow scientists. It has never bothered me when I see my books shelved in the self-help section. Such categorizations are complex, blurry, and arbitrary.[6]

Learning how to read, and studying math, history, art, and sports, gives us resources that allow each person to freely find their own path. The same thing happens when we apply that idea to self-knowledge. Philosophy itself has oscillated over the centuries between skepticism over its ability to offer practical knowledge (such as in the legendary discursive battles Socrates waged with the Sophists), and other periods where its main objective was to offer advice on how to lead a virtuous life, which was what Cicero and Marcus Aurelius devoted themselves to.

This book similarly oscillates. It is a scientific account of the human mind in which ideas for improving our mental and emotional lives naturally crop up. I think it is worth distilling them into an action-oriented chapter summary. These are food for thought, not formulas; I don't believe magic formulas exist—in any realm of knowledge—that can transform us without real effort. There is no book that will make us good tennis players, or good chemists, or good industrial engineers. Nor is there any manual that can turn us into good people just by reading it.

That said, I am confident that these ideas will be useful to someone. They do not claim to be universal. Some will seem far-fetched. Others will strike a chord and hopefully serve as the starting point for a practice that can contribute to a better life.

---

6 The worst case of bibliographic classification I've ever seen was at a bookstore in La Plata, Argentina, where they'd decided to put *Alive* in the cookbook section.

1. **Measure the words you use to refer to yourself**
   The words we use to describe how we feel have—in and of themselves—the power to influence our moods, becoming self-fulfilling prophecies. It is worth making an effort to use them precisely, noticing the nuances. Maybe, instead of feeling "horrible," you are just sleepy or hungry.
2. **Remember that sometimes (in fact, often) you're wrong**
   Don't subscribe fully to first impressions even when evaluating your own mood. Are there alternative explanations? Other ways to look at it? Important details to take into consideration? Your first impression is just that: a hypothesis that can be improved upon, or even changed completely.
3. **Gain perspective, look at yourself with more distance**
   We are often the objects of our harshest judgments. Try the exercise of considering your case while imagining it is happening to someone else, with dispassionate distance, from a place where things don't seem as catastrophic or serious.
4. **Dialogue helps us to think**
   Talking to other people clarifies ideas, and helps us to find errors in our own reasoning and identify better solutions. It also helps us to learn to converse better with ourselves. It is our most powerful tool for improved thinking.
5. **Conversation only works in its natural habitat**
   Not every conversation has the same value. It is only conversations in small groups of people, who are receptive and predisposed to being swayed, that are effective. In other words, we refer here to dialogue in good faith toward a mutual process of discovery.
6. **Mass public conversation doesn't have the same effect**
   Social networks have a particular dynamic and inertia that do not facilitate conversation. They foster a type of discussion where it becomes very difficult to engage in a

constructive exchange of viewpoints and to articulate any consensus. They often only manage to provoke tension and make us double down on our positions.

7. **Relativize**
   Surely this has happened to you before: something that seemed unbearable at the time now seems trivial or at least less important. Our irrefutable explanations of today may seem absurdly exaggerated in the future. Remember that, especially when you're intoxicated by an emotion.

Answers to the problems on pages 22–3. What smells just like white paint and is red, is red paint. If there are seven apples and you take two, you have two. Not five, as our automatic system impulsively answers, thinking of the subtraction rules we heard so many times at school. The same trap affects the problem of the race: if you pass the second-place car in a race, you are in second place, not first as most people would think. If a white horse goes into the Black Sea, he comes out . . . wet. The color of the horse is irrelevant and distracts from the evidence needed to solve the problem (the horse comes out of the water), making it less available. The solutions are all obvious, but often we don't even consider them. The way to go through all nine dots with four lines is hard to find because we don't even think of the key to the solution: the lines don't need to be contained within the square.

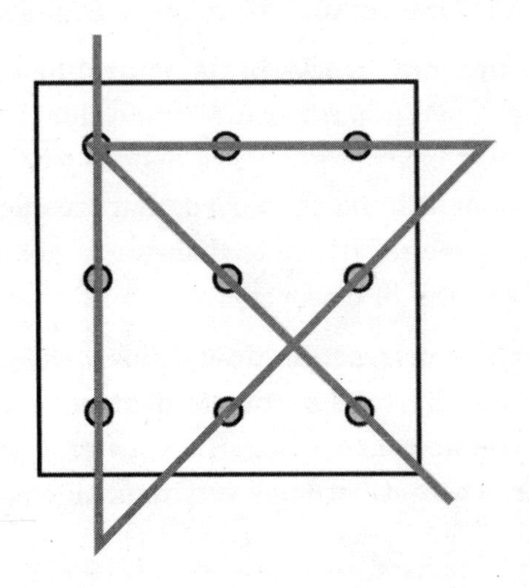

CHAPTER 2

The Art of Conversation

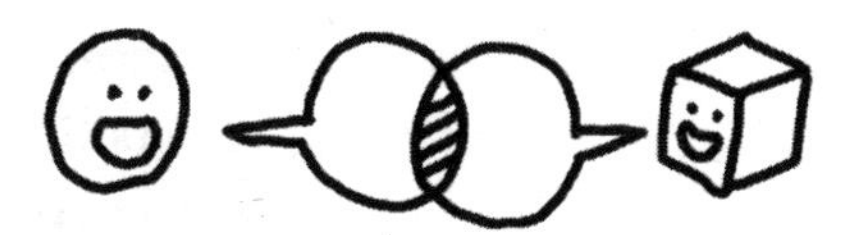

How to make better decisions

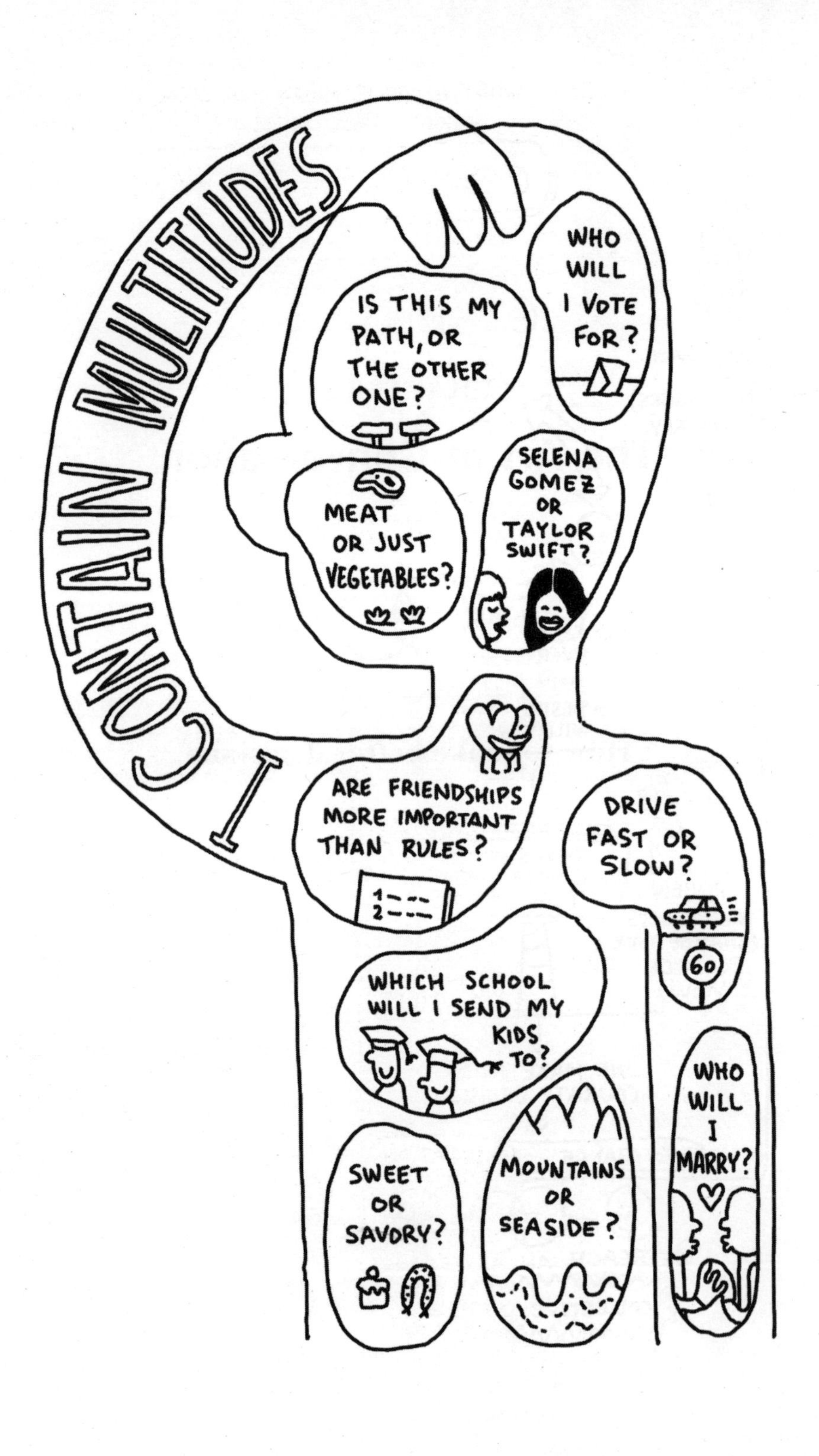
I CONTAIN MULTITUDES
IS THIS MY PATH, OR THE OTHER ONE?
WHO WILL I VOTE FOR?
MEAT OR JUST VEGETABLES?
SELENA GOMEZ OR TAYLOR SWIFT?
ARE FRIENDSHIPS MORE IMPORTANT THAN RULES?
DRIVE FAST OR SLOW?
60
WHICH SCHOOL WILL I SEND MY KIDS TO?
SWEET OR SAVORY?
MOUNTAINS OR SEASIDE?
WHO WILL I MARRY?

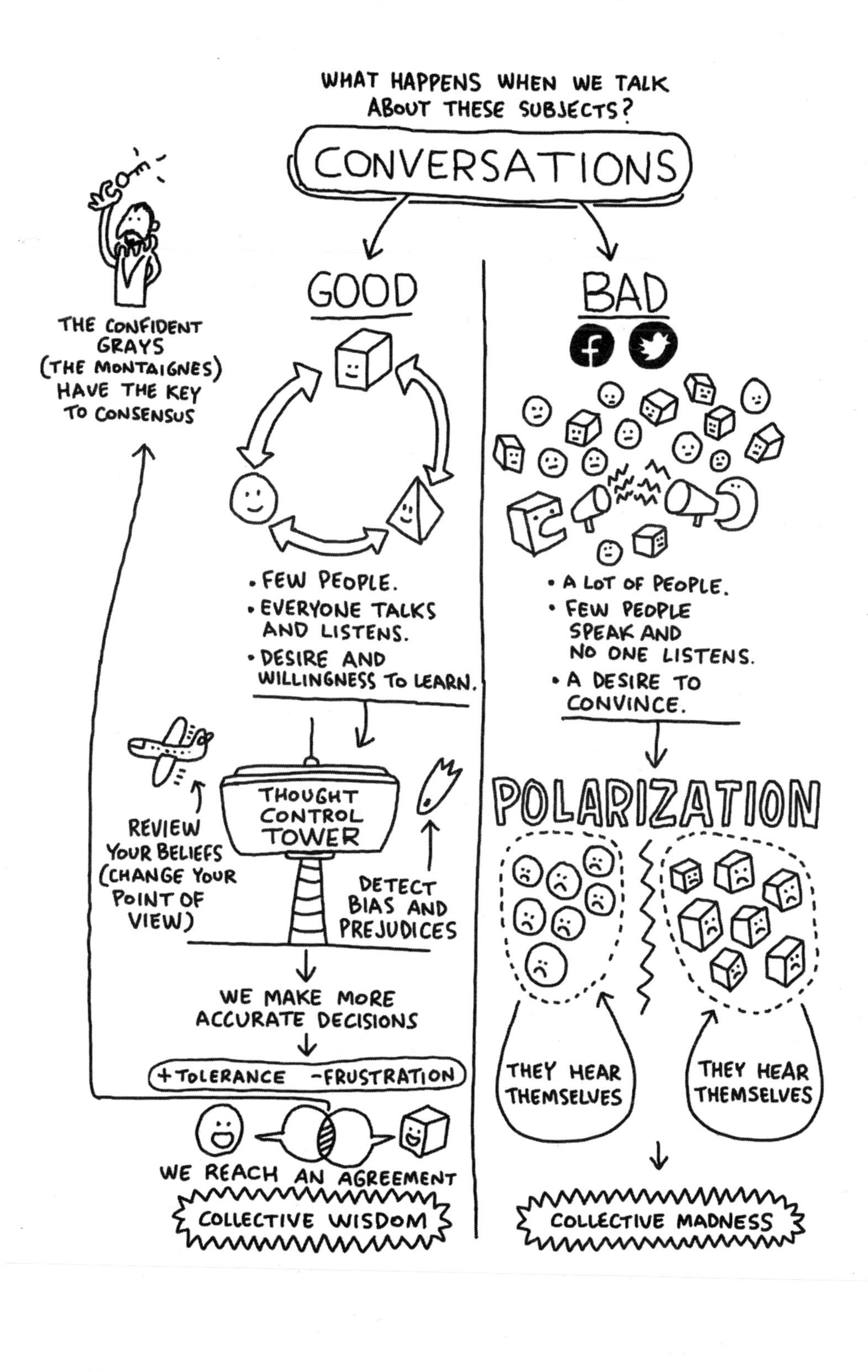

WHAT HAPPENS WHEN WE TALK ABOUT THESE SUBJECTS?
CONVERSATIONS
GOOD
BAD
• FEW PEOPLE.
• EVERYONE TALKS AND LISTENS.
• DESIRE AND WILLINGNESS TO LEARN.
• A LOT OF PEOPLE.
• FEW PEOPLE SPEAK AND NO ONE LISTENS.
• A DESIRE TO CONVINCE.
THE CONFIDENT GRAYS (THE MONTAIGNES) HAVE THE KEY TO CONSENSUS
THOUGHT CONTROL TOWER
REVIEW YOUR BELIEFS (CHANGE YOUR POINT OF VIEW)
DETECT BIAS AND PREJUDICES
WE MAKE MORE ACCURATE DECISIONS
+TOLERANCE -FRUSTRATION
WE REACH AN AGREEMENT
COLLECTIVE WISDOM
POLARIZATION
THEY HEAR THEMSELVES
THEY HEAR THEMSELVES
COLLECTIVE MADNESS

Our most frequent decisions often don't have an exact solution. They include social decisions (how and with whom we associate); political decisions (who we vote for and what ideals we sympathize with); our general vision of the world, and our morality. These last two are of particular interest because, although they are the most unquestionable convictions we have, most of the time we don't know how they were formed or what might make them change.

Should we take this path or another one? We base our decision—as is almost always the case—on a combination of intuition, chance, and ignorance. Talking about these decisions in the right context greatly improves their precision. On the other hand, talking produces the opposite effect when there are too many people involved, when someone monopolizes the conversation, or when a spirit of confrontation prevails. In the latter type of conversations, our errors become entrenched and mistaken ideas can be vigorously reaffirmed. We need to identify the fine line that divides collective wisdom from collective madness. This takes on practical relevance: many organizational catastrophes are simply the result of not having created the appropriate atmosphere for good conversations.

On November 8, 1960, John F. Kennedy defeated Nixon by a slim margin and became the youngest president-elect in the history of the United States. In one of his first meetings, the military chiefs of staff informed him that Assault Brigade 2506, made up of 1,500 Cuban exiles trained at a CIA base in Guatemala, was ready to attack the Cuban coasts and overthrow Fidel Castro's regime. In those dramatic conversations welcoming Kennedy into office, his military advisers emphatically recommend continuing with the plan. Their opinion is *almost* unanimous. The only opposition comes from Arthur Schlesinger, his campaign adviser; in a memorandum he expresses the numerous reasons why he considers the attack unwise. Intimidated by the conviction of the rest of the council in the decisive conversations, Schlesinger stands by and remains silent. Kennedy hears only one position and on April 15, 1961 he launches the attack from Nicaragua. The following day, Fidel Castro gives one of his most famous speeches, before a crowd armed with crude rifles, and he calls on his people to defend the nation. Cuba repels the invasion in just four days.

So Kennedy began his presidency with a resounding defeat. And a great lesson. Once the crisis was past, the American administration completely changed its decision-making protocol. Specifically, JFK broke up the advisory meetings into smaller groups in order to ensure that all opinions would be expressed and heard. And, with the objective of not silencing any voices, he and his generals would often abstain from attending the meetings. He also created a key mediator figure, his brother Robert Kennedy, who led the talks from a more even-handed and balanced position.

Just two years later, in October of 1962, this new way of deliberating is put to the test in the second Cuban crisis, which could have tragic consequences for the entire planet. The US Secret Service detects Soviet missiles in Cuba that have nuclear capabilities. Tensions are sky-high and growing until, a few days later, we are the closest we've been to a Third World War. The military advisers again push for an immediate attack, but this time not with a troop poorly trained in Guatemala, but with nuclear fury and an arsenal unprecedented in the history of humanity. Kennedy's advisers are divided. Another group favors establishing a naval embargo that halts the flow of Soviet arms to Cuba to buy some time and make it possible to negotiate a peace accord. Every voice has its chance to speak at the council meeting and they even prepare two different presidential addresses at the same time so that the decision, once it has been made, can be communicated as soon as possible. Kennedy listens to all the opinions, the arguments on both sides, and with that partial information he makes one of the most important decisions in human history. Finally, he opts for negotiation, and the conflict is resolved in a matter of days.

Between the madness and wisdom of crowds

Kennedy's decision in the Cuban missile crisis has had unique historical consequences, but it also shares countless similarities with other decisions we make all the time. Firstly, because the procedures are not that different and secondly because, while our decisions don't usually change the fate of the planet, they do change our fates. And from our perspective they sometimes take on vast proportions.

When choosing a school for our children we take into consideration various factors that are difficult to compare: distance, quality of the education, the social and emotional environment, the economic cost, and, sometimes, even ideological and historical questions. All these arguments compete and the decision is settled in a debate we rarely consciously recall. The same happens when we vote, when we choose our vacation spot or what dish to eat from the long menu we've been given in a restaurant, and when we define our social and romantic life.

It is in this field, which is much more obscure and random than logical, that most of our issues lie. We make decisions with a partial view from which it is impossible to predict the exact consequences of each option. Intuition and hunches spontaneously play a role; rather than us making the decisions, the decisions make us.

Those decisions that are made intuitively are perceived very differently from those based on reasoning. It seems that they take place not so much in our brains as in the rest of our bodies, and that is the source of the metaphors used to describe them: a gut feeling, smell out a situation, have a thirst for vengeance, hunger for glory . . . But it turns out that reason and intuition aren't actually that different; the former is conscious deliberation, the latter unconscious deliberation. When we feel a strong intuition, it's because the brain is analyzing, unbeknownst to us, our options and their possible consequences, and then expressing through our bodies which is the preferred choice.

The impossibility of perceiving our "unconscious reasoning" gives intuition its mysterious aspect. When schoolchildren start to add and subtract, they usually hide their fingers because it is more fun to solve a problem without revealing the working mechanisms behind it. In much the same way, our gut feelings are briefly hidden. How would we feel if we could see all of the images projected into the future the way they are in our unconscious mind? In the Marvel universe there is a hero who caricatures this experience: Dr. Stephen Strange, who has the ability to simulate, in the animated theater of his brain, the futures of millions of alternatives. This caricature allows us to see what is happening in the opacity of the unconscious when we consider the option that is closest to what we desire. In a way, we've already discussed this idea: it is that multitude that we each contain.

The fact that the "logic" of intuitive decisions is not visible to us makes the practice of conversation even more necessary. By bringing those arguments to the surface, we can identify biases, errors, and priorities which would otherwise be impossible for us to see.

Conversations over coffee

There are a few stops we must take along the path that joins Mercier's logic with the governing of our own minds. Let's begin with simple problems that simulate making decisions with partial information: What is the surface area of Australia? How many people live in Potosí? What is the height of the Eiffel Tower? How many buses circulate in a day in New York? How many kisses, on average, does a person give over their lifetime? Despite there being a correct answer for each question, they are practically impossible to calculate precisely. Many of the questions are part of the board game El Erudito, in which players have to make an estimation and the one who's closest wins. It is like Trivial Pursuit, but rather than requiring perfect answers, it rewards approximation and comparative thinking.

Joaquín Navajas, Gerry Garbulsky, and I chose some of the questions in that game for an experiment designed to determine whether conversation improves intuitive and approximated decision-making. The experiment was quite unusual because it was carried out in a theater with ten thousand people, in perhaps the largest ever simultaneous public debate.

We had fifteen minutes to get thousands of people playing in such a way that they each discovered something about themselves. Something special occurred that day. Sometimes, unexpectedly, science happens.[7] Onstage, under time pressure, leading a live experiment with ten thousand people, I remembered my days as a student in New York, when I was recording the neuronal activity of the brain's visual cortex. In the middle of that microscopic world of brain tissue, which was brand-new to me, what most surprised me was the amplified noise of the currents of ions entering cells. Years later, surrounded by the commotion of thousands of simultaneous conversations, I was again listening to the soundtrack of an experiment. Then all my accumulated nervousness calmed and I understood that, beyond what the data would reveal, the experiment was already a success. We left with bags filled with pieces of paper, each with a handwritten response that reflected the beliefs, hesitations, and certainties of the ten thousand people gathered in the stadium.

The experiment went like this: from the stage we asked eight questions and each participant jotted down their answers and how confident they were about each one. The next step was to make groups of five people. Those groups would discuss the questions for three minutes and together come up with the best possible answer. I've done this experiment on all sorts of different stages, from schools to big companies and financial institutions. I'm always surprised by the energy, competitiveness, and determination that each group brings to winning a game where there is no prize except for the satisfaction of feeling you've reasoned correctly. It is the pleasure of discovering the truth, which turns out to be a fantastic motivation. Once they've finished their collective reflection, each participant goes back to their seat and can see whether they've changed their mind.

---

7 Not as often as shit, unfortunately.

To illustrate the results, imagine a simpler situation involving only two people. Responding to the question about the height of the Eiffel Tower, one of them replies that it's three hundred meters tall and rates their answer with a confidence of nine out of ten; the other person says that its height is two hundred meters and gives his reply a confidence rating of three. The average of the two answers is two hundred and fifty meters. In an average weighted toward the more confident person, however, the answer is closer to three hundred meters. This is the optimal way to combine group results. It can be mathematically proven that this procedure is the one that yields the response that is closest to correct.

Analyzing the answers of thousands of groups, we discovered that this was the most frequently used strategy. Without any instructions, without having agreed upon anything, without knowing each other and in just a few minutes, the groups found out how to arrive at the best possible answer: that's how effective good conversation is.

These results can be compared with a criterion established more than a century ago. In 1907, Sir Francis Galton asked 887 people (nonprofessionals) to guess the weight of an ox. He discovered that the average of their opinions was more accurate than the estimation of the greatest experts and that led him to coin the famous term "the wisdom of crowds." It works through a very simple statistical principle: everyone makes mistakes, but if you average out all the opinions those errors cancel each other out. This is not exclusive to people, ideas, or opinions. As repetitions increase in any "noisy" system, its randomness and fluctuations vanish. We see this idea in television game shows when the contestants ask for help from the audience. They are consulting the wisdom of crowds.

A little over a hundred years later we find that small groups are even wiser than crowds. The average weighted for confidence is better than the simple "Galton-style" average. But there's more. Groups do something even more effective: they rethink the problem together, explain to each other how they've each arrived at a conclusion and review those procedures, which substantially improves their estimates. We observe the same process in Mercier's experiment: conversation makes visible the mistakes that lead us to make bad decisions.

And what can we do if it turns out we can't have a conversation before making a decision? A good suggestion is found in professional chess: the players use part of their allotted time to write down their move before making it. The moment they put something in writing, it turns out, they can detect a mistake they've overlooked. In that territory so limited by rules, writing down a move is the closest thing to having a conversation. And this also applies, in a way, to all our important decisions. Before making them, we should write them down. It's even better if we can tell them to another person. This is the only way we will discover the overlooked errors in our reasoning.

The madness of crowds

The article where we showed that a brief conversation significantly improved decision-making caused a stir in the scientific community because it contradicted a well-established belief. Charles Mackay, a

THE QUICKEST ROUTE IS THIS ONE,
I'M A HUNDRED PERCENT POSITIVE:

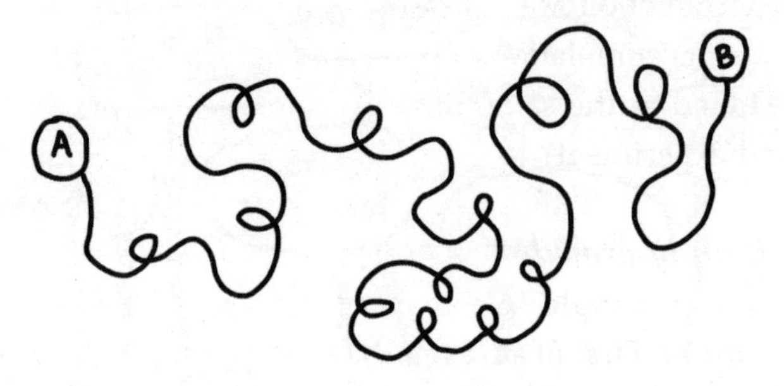

WHAT DO YOU THINK?

Scottish journalist, collected a large number of historical events in the book *Extraordinary Popular Delusions and the Madness of Crowds* (1841), in which the fervor of mass dialogue had helped to spread, like a virus, some extraordinarily delusional ideas.

Thirty years before the telephone was invented, Mackay already understood that there was a common thread in all those events: when discussions take place in large groups, certain viewpoints spread and infect individual opinion, causing the crowd to lose the trait that previously made it valuable: its diversity. The errors (of over- and underestimation) become common and, instead of canceling each other out, they are enhanced. That is how the popular delusions Mackay describes get their start, through contagion. Ideas, like laughter, tears, fear, and enthusiasm, are highly infectious.

The list of historical delusions includes the witch hunts, the Crusades, wars . . . These phenomena share the reflexivity we saw in the financial bubbles. Crowds, which today gather more than ever because of social media, have a strong tendency toward delusion. It is almost a hallmark of our era. Mackay would have had a field day with Twitter, or X as it is now called.

The contradiction we must resolve is the following: how is it possible that accumulation of opinions leads to delusion, as Mackay suggested, and at the same time in sound judgment, as we saw in our crowd experiment?

The wisdom of crowds

The answer is simple. *Good conversations only occur in their natural habitat.* First of all, even though it sounds basic, the groups have to be small. In a crowd people fire off verbal shots, but they don't converse. There is neither the time nor the desire: those shouting want to be heard, not to listen to others. Secondly, people have to have an open mind and be predisposed to listen and exchange ideas.

This principle was already well known to the ancient Greeks, who were pioneers of constructing a shared vision of the world through conversation. Philosophy, as Plato pointed out in *The Symposium*, comes out of conversation and not, as we imagine it today, out of writing in an isolated room. The Socratic symposium included a tragic poet, a doctor, and a comic poet, people with different perspectives who stretched out comfortably to share some food and drink with music playing in the background. That was the ideal context for trading ideas through conversation. And that is where the word symposium comes from, although today we use it to mean a conference given by experts, but its etymology is: *sun* (together) plus *potēs* (drinker). In other words, a symposium is an excellent setting for conversation: *drinking together*.

This ode to good conversation is repeated cyclically throughout human history, with high and low tides of symposia, banquets, and good spaces for sharing ideas in words. Some five hundred years ago, Michel de Montaigne anticipated the Enlightenment and humanism with this same premise: he pointed out that the habit of good conversation as a laboratory for ideas was being lost. And so, in his essays, he outlined the principles of *the art of conversation*:

- Don't get offended when someone thinks differently, and embrace those who contradict you.
- Don't speak to convince, but rather for pleasure. Appreciate the exercise of reasoning.
- Speak in your own words, don't repeat quotes like an encyclopedia.
- Doubt yourself; remember that we can always be wrong.
- Use conversation as an essential space for judging our own ideas.
- Value ideas merely for the impact they have when we put them into practice, just as we respect a surgeon for his operations and a musician for his concert.
- Think critically.
- Do not mistake beauty for truth.
- Avoid prejudices, carefully distinguishing concrete examples from generalizations.
- Organize your ideas well and carefully review your arguments.
- Reflect on what you've learned from others in the conversation.

Montaigne is a conversational hero, an atypical hero who, despite not being the strongest or the fastest, understood that words are the most virtuous tool for shaping our ideas, and he used this knowledge to resolve one of the most violent conflicts of his time.

We take up his ideas again—ideas that never lost currency with the great thinkers—and we turn them into science. Our conversational space is that of a crowd of small groups, not a huge crowd. This is the key, just as Kennedy identified after the categorical defeat at the Bay of Pigs, when he decided to divide his large advisory board into small discussion groups.

In our experiment, each of the ten thousand participants returned to their seats after conversing with their group and, once there, chose whether or not to change their opinion. The vast majority did shift their answer, arriving at a much better conclusion than both their original one and the group's. The analysis of these responses reveals the risks (by contagion) and the benefits (by revision). Let's begin with the riskier side. After wrapping up a conversation, the opinions of the members of every group are more similar. The richness of diversity is lost due to the influence some people exert over others. But the contagion is not widespread because being separated into groups acts as a firewall for collective delusion. The benefits, on the other hand, are much more prominent. The opinion shifts produced after talking are almost always shifts in the right direction.

In short, when ordering the answers by their degree of accuracy, this is the result: the worst are those that people make before talking; then come the ones obtained by averaging all the opinions (Galton); even better are those derived from a confidence-weighted average (the optimum algorithm for combining results); the answers resolved in a group, with a discussion of the arguments and procedures, are noticeably better; and the opinions each individual forms after this discussion are remarkably better.

The conversations that take place in small groups retain the best of both worlds: on one hand, the process of reviewing and correcting mistakes, which are only noticed through dialogue; on the other hand, since the groups are small, they allow a degree of statistical independence thanks to which the crowd doesn't form a monolithic opinion block. This is the point at which the wisdom of the crowd meets the madness of the crowd—the territory where dialogue is most productive.

Fear in the meeting, fire in the hallways

Around a large table, whether the setting is a business meeting or a political gathering or a family meal, there is always one person who monopolizes the conversation. It may be that they are the hierarchical leader, or just the most extroverted. Either way, what happens is that most people listen, and despite the fact that many disagree, they don't say anything, because they're too shy or intimidated. Once the meeting is over, chatting out in the hallway, those same people speak their minds with those closest to them. Only there do all the problems emerge. A great opportunity is missed because of the size and shape of the meeting table.

Margaret Heffernan tells how this has led to great disasters. On October 29, 2018, Lion Air flight 610 crashed shortly after takeoff from Jakarta airport. A few months later, on March 10, 2019, there was another very similar accident during a takeoff in Addis Ababa. Three hundred and forty-six people died.

Why weren't the errors identified shortly after the first accident in exhaustive reports by experts? An essential aspect of the problem was the lack of appropriate areas where the risks could be identified and communicated; silenced by the fear that prevails around the meeting room table, people poured out their feelings in the hallways.

This problem is common to every organization, and the larger and more complex the organization, the bigger a problem it is. On many occasions it can be remedied without spectacular interventions or great technological fanfare, by merely replacing ineffectual dialogue that reaffirms the status quo with effective conversations.

The blurry outline of the acceptable

Let's deal a final blow to the conversation crisis, by hitting it where it seems to fail most utterly: in the moral and political realm; in areas of polarized beliefs that seem to have no room for maneuvering. We will show once again that, even in that chaos, words have an extraordinary capacity for bringing about transformation.

A year after asking people to debate the height of the Eiffel Tower, we returned to the same forum to up the ante, to repeat the game but with much more complex and controversial questions about topics such as abortion, sacrificing lives so that others can live, prioritizing friendship over lawfulness, and genetic engineering. Obviously, I have an opinion on these issues, but that is not important here. What is important is sparking an open debate without prejudices that allows us to review and understand our stances and opinions, even on taboo subjects where dialogue is more difficult.

In preparation for this experiment, we used surveys to measure which problems are most polarizing and then we calibrated each one so that the divide between those for and against was equal. Each scenario involved a particular action. For example, one scenario presented siblings, both adults, who loved each other and, in an informed and consensual decision, had sex. Each group had to respond to the question of whether what they'd done was acceptable. The answer was on a scale of zero to ten; zero being completely unacceptable and ten perfectly admissible.

One consideration that arose with the sibling lovers scenario was whether they might conceive a child with some sort of genetic malformation. This is interesting, because we had never mentioned the gender of the siblings or the sort of sexual relations they were having. So there was an assumption that it was a heterosexual relationship with vaginal penetration and a possibility of pregnancy. We thought that eliminating that dilemma would make it more acceptable and we presented a different scenario: the siblings decided to have oral sex. However, this limitation made it more easily visualized and that provoked more intense physical reactions. As a result, on average, people rated that version of the scenario even more unacceptable.

I almost cried at the sight of ten thousand people debating uncomfortable issues. It was an exercise in science, but also in freedom. I keep photos of that day as a vivid record of the emotions that arise in these sorts of conversations. I'm particularly fond of one that we took of a group discussing a classic moral dilemma styled after *Inglourious Basterds*, the Quentin Tarantino film: a family hides in a basement after a military invasion; if they make any noise, they'll be found and all killed, including a newborn baby in his mother's arms. Inevitably, the baby starts crying. His mother desperately covers his mouth. She tries everything to calm him, but nothing works. At some point she realizes that the only way to save the rest of the family is to sacrifice the baby, and she decides to kill him. The question is: is what she did acceptable? In an impromptu ethics commission spread out across the stadium, thousands of groups debate in unison. In one of them, a woman argues fervently with a baby in her arms.

We usually have fairly clear opinions about such dilemmas. Furthermore, those opinions feel categorical, absolute. We don't generally see degrees or quantities, and we have the impression that we must take a firm stance, on one side of the line or the other. It doesn't actually work that way. First of all, any problem can be subtly tuned until nearly everyone changes opinion. Let's look at an example. When people find themselves in the situation of having to report a friend who's committed an infraction, it usually comes down to a matter of principles: on one side are those who prioritize the law, and on the other those who value friendship above all else. Germans and Argentines. Now, if we gradually increase the severity of the infraction, we will arrive at a turning point where the two extremes even out. Once that point is passed, the balance tips in the other direction.

What is even more surprising is our ability to change our minds about opinions that seem so deeply rooted. In that stadium packed with conversations we discovered that *our beliefs are much more malleable than we think.*

The confident grays

What is the likelihood that people with completely opposing opinions on thorny moral and ideological issues can reach an agreement after a few minutes of talking? Collective supposition, informed by social media interactions, is that the probability of reaching consensus is infinitesimal, another reflection of the great skepticism of dialogue's capacity for smoothing out differences.

Is that pessimism justified? We respond to this question by contrasting collective supposition with the experiments we carried out in Buenos Aires and Vancouver, on the main TED stage. Each person determined, on a scale of zero to ten, how acceptable they found each situation. Then they gathered in groups of three—like an ethics commission—to try to reach consensus: a number that would sum up the group opinion on the dilemma's acceptability. The decision had to be unanimous among the three members of the group. As in Fonda's trial, one dissent was enough to invalidate the consensus.

As was to be expected, the probability of reaching consensus decreased when the opinions of the group members were diverse. What is most noteworthy is that the probability of reaching an agreement in groups whose participants had completely antithetical views was between 30 and 50 percent, depending on the question and the location (to my surprise, the rates of consensus were higher in Buenos Aires than in Vancouver). In any case, with every question and in both places, the likelihood of consensus was much higher than people estimated. Reality turns out to be much more prudent, flexible, and open than we imagine.

The next step was to figure out why some groups were so much more likely to reach consensus than others when starting from the same antagonistic position. The key is found in some atypical characters: the confident grays. Let's see how that works.

People with extreme opinions tend to be very confident about their answers. On the other hand, the "grays"—those who've deemed the dilemma acceptable to some degree—are more hesitant. This rule reveals something quite fascinating. There is a small group of people who respond with intermediate degrees of acceptability, but with great confidence in their answers. They are gray because they are convinced that the moral dilemma presents good arguments on both sides, even when that results in a contradiction. We found that *confident grays are the key to consensus*; they are the ones who can bring together two people with opposing ideas.

Here again appears Michel de Montaigne, the patron saint of free thinkers. In his essays, Montaigne created a way of thinking about any subject, no matter what doubts or reluctance you have. Any and every idea can be thought about and discussed. This was not pure rhetoric. Montaigne flung open his doors and hospitably received, with banquets and conversation, those who threatened him, and that saved his skin more than once. And, as is often the case with confident grays, Montaigne was bludgeoned by those on both sides of the trench that separated Catholics and Huguenots in his day. Conversation was always his defensive weapon and his way of mediating the conflict, so much so that he became a decisive figure in the promulgation of the Edict of Nantes, in which King Henry IV granted religious freedom in France and put an end, at least for a while, to that long period of bloodshed. Montaigne is *the* quintessential confident gray. From here on, I will call the confident grays the Montaignes, in homage to his legacy.

The dilemmas in our experiments are simple caricatures of moral thought: controlled, bite-sized, almost mathematical situations. This design is ideal for experiments, but it makes them less realistic. Perhaps the power of conversation, from Mercier's logic to intuitive and moral decisions, works in the lab, but not in "real life," where everything is more complex. Therefore, we will take one more stab at it,[8] and leave the laboratory to make sure that conversation continues to work outside of it. We will do this by journeying first back into childhood, where our understanding of the world is tacitly formed, and then to Jerusalem, the site of one of the most unresolvable conflicts in recent history.

---

8 The last one is never really the last. A famous example is the interminability of a round of beers; hence the existence of terms that refer to the last of the last drinks: the *saideira* in Brazil, the *espuela* in Spain, *la del estribo* in Mexico, and what in English is called *one for the road.*

The roundness of the Earth

Messi or Cristiano? Legal abortion? In Spain, separatism or constitutionalism? In Argentina, Peronism or anti-Peronism? Trump? Each of these thorny issues opens up an ideological rift in its specific terrain. There is one in particular that is of interest because of the bizarreness of the intellectual debate it raises: the rift that separates those who are convinced the Earth is round from those who believe it is flat. This discussion, apparently laid to rest several times in history, has two iconic images. The first is of Eratosthenes measuring the circumference of the Earth—with surprising precision—from the difference in the shadows cast by the sun, at the same time, in Syene (modern Aswan) and Alexandria; the second is of Christopher Columbus proposing to a stunned Queen Isabella that he can reach the Indies by sailing west. The chronology of these stories is striking. Eratosthenes discovered and measured the Earth's circumference about seventeen hundred years before Columbus unwittingly set sail for the Americas. The most common way that the two stories have been reconciled is that human culture entered some sort of black hole during the Middle Ages, when all Greek knowledge was confined to libraries and a chosen few were responsible for keeping the flame alive until the Renaissance. According to Jeffrey Russell's *Inventing the Flat Earth*, the story of Columbus's audacity is nothing more than a myth that's spread like wildfire, as tends to happen. Russell writes that the idea of the spherical Earth was widely accepted already by the Middle Ages and certainly during the time of Columbus and Isabella.

Had Russell waited a little longer to publish his book, some doubts probably would have arisen in his mind about the unflagging endurance of this idea that had already been proven. In 2017, just sixteen years after his book appeared, the first Flat Earth International Conference was held in North Carolina, giving rise to the movement of flat earthers.

Our ancestors' difficulties in conceiving of the Earth's roundness are the same as those manifested in each individual's cognitive development. There is a parallel between the conceptual revolutions in a child's learning and those in the history of culture. In early childhood, our senses apprehend the flatness of the Earth as something evident, and such a natural conclusion is difficult to debunk. How can a five-year-old understand that the planet is round when it looks flat in every direction? How can she grasp that it floats in space, with nothing holding it up? If it's round, how come the people on the other side, on the bottom, don't just fall off? How can it be that it is round and those below, on the other side of the world, do not fall? The great conceptual revolutions in cognitive development involve building systems that are irreconcilable with those they replace. And this leads us—happy coincidence—to the land of Eratosthenes, at the modern University of Athens, where Stella Vosniadou, a professor of cognitive neuroscience, made the most thorough and exhaustive study to date on the years of transition between a child's conception of the world as necessarily flat and her understanding that it is round, with everything that such a change in worldview entails.

It turns out that this transition is much less immediate and much more costly and contradictory than we remember. Carefully analyzing this spontaneous process that revolutionizes our thinking gives us helpful clues for getting out of a rut as adults.

What does it mean to a child's mind that the world is round? To reconstruct this mental representation, Vosniadou proposes a series of questions such as: where does the Earth end? From what vantage point can the Earth be seen as round? Why do we always see the stars, moon and sun by looking up? Children respond to these questions with words and drawings that reveal what they really mean when they say that the world is spherical. Vosniadou's findings are that almost all children go through three phases, each of which is represented here by drawings made by students at the Liceo Jean Mermoz in Buenos Aires, where Diego de la Hera, Cecilia Calero, and I carried out an experiment similar to Vosniadou's.

1. FLAT MODEL

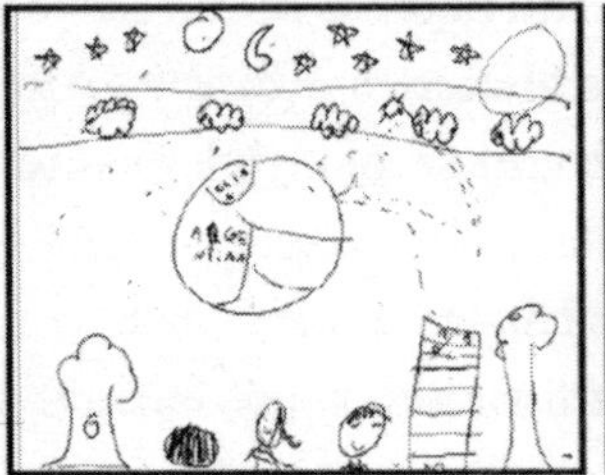

2. DUAL MODEL

3. HOLLOW MODEL

1. The Earth is round. But it's a disc, not a sphere. This is the most natural and simple way to reconcile what they used to believe (which is that the Earth is a flat surface) with what they have just learned (that it is round).
2. In this model the Earth is already a sphere, it floats in space, it has seas, you can see America and China. The Earth looks like the globes they've seen. But, in their minds, that isn't where we live. In this mental model, the Earth is like the moon or the sun, one of the many objects in the cosmos. There is another, separate plane we inhabit.
3. The Earth is round. At this stage, they understand that this refers to the curvature of the surface. But they still can't grasp why people on the bottom don't fall off. Their solution is to draw a concave Earth: we live in a bowl.

None of these models corresponds to reality, but they are all brilliant, surprising creations of the imagination's attempts to accommodate, as best it can, all the data the children have at their disposal. They represent a way of making what they've been told concur with what they experience. The dual model has always seemed to me to be an example of the extraordinary virtuosity of human thought, of our ability to fit together seemingly incompatible pieces into a simple, elegant, and creative solution.

Diego, Cecilia, and I brought together children with different mental models of our planet to work in pairs on drawing the Earth, the clouds, the sun, their house, the other side of the world (China), the stars, the sea . . . The project required consensus; two children with different opinions about the world were trying to agree on how to depict it. Then, each child went back to their own spot and solved the exercises that Stella Vosniadou had used to classify the mental models during the conceptual shift.

We found that, within almost all pairs, those with the weakest arguments tended to change their representation and those whose arguments were more consistent with reality did not revert to a more precarious model. In other words, the thinking the team agrees upon does not arise out of any sort of social leadership, nor some exercise of concessions and mutual negotiations to converge at an average; the most convincing arguments simply prevailed. Just as Fonda turned the twelve men of the jury. Among children or adults, in small groups, good reasoning has decisive weight.

Without this activity, all the children would have gradually, through several cognitive leaps, shifted their representation toward a more suitable model. It turns out, then, that a simple conversation, a joint game between peers, is capable of precipitating and enormously accelerating this conceptual revolution. Peer conversation is an extraordinarily powerful and effective tool in schools. First, because it catalyzes the development of ideas, and second, because it makes us better at an essential art: using words well.

A Jewish Moor

Let us now return to the adult world, where these conceptual battles can and often do have much more violent overtones. For that very reason, it is even more urgent to find a way to improve them. Let's travel to the epicenter of one of these battles, Jerusalem, where nations, peoples, religions, and an endless history of wars and conflicts converge in an area just over a hundred square kilometers.

I traveled to Israel for the first and only time in 2012, on a flight from Madrid to Tel Aviv. I rushed to the boarding gate only to find that another security check was still required. Border crossings always scare me; I can't help but worry about the myriad papers, things in my pockets, or arguments I may have forgotten. I always feel intimidated, nervous that I've broken some regulation. At the same time, I am hyperaware of how lucky I am, of how many people truly risk their lives at border crossings.

Mine was very simple. I was going to a conference at the Hebrew University of Jerusalem in Ein Gedi, a beautiful place on the shore of the Dead Sea. This time I went to the security checkpoint in such a hurry that I didn't even have time for nerves. I assuredly answered all the questions, until the officer asked me whether I was Jewish, and I realized that, just as with the moral dilemmas in our experiment, the answer was not binary. So I began a process of deliberation, at top speed, to figure out who I am.

I come from a Jewish family; I feel a strong bond with Jewish culture, food, and music. Occasionally I've celebrated some Jewish holidays, but so irregularly that I have almost zero knowledge of the rituals, traditions, history, and dates. Sometimes I even mix up their names. I've never gone to synagogue, I'm not circumcised—it will never cease to amaze me that penis morphology is a religious covenant—and I never went to Jewish schools or clubs. But when my grandfather gave me Howard Fast's book *My Glorious Brothers* and wrote in the dedication, "So that you always love the Jewish people as well," I felt the story of the Maccabees as if it were my own. And, over the years, I've also found that I can't escape being Jewish, because that's how others see me. I discovered that in hostility and aggression, when I was newly arrived in Argentina and I found my classmates had placed on my desk a bar of soap with my last name written on it. To them, I was human soap; they identified with the role of the soap makers. So, in a way, I discovered my Jewishness when it became clear that others saw me as a worthless Jew.

All these thoughts went through my head in the few seconds—which the officers obviously thought were too many—it took me to come up with an answer: "Kinda." It wasn't a joke, or an attempt to provoke them. It was the most succinct and honest answer, and also the one I thought would get me boarded on time. A huge mistake. Soon I found myself in a basement with the other confident and not-so-confident grays: those who didn't fit into any of the preestablished categories. I was asked about all the elements of Jewish identity, including many of those I'd outlined during my attempt at an answer: my family, the schools I went to and, of course, the morphology of my penis.[9]

Boarding that plane was just the prelude to one of the most intense trips of my life. Every minute and every place was charged with strong emotions, from the geographical and political ambiguities of the road to Masada, the limitless Tel Aviv nights, the infinite calm of the sunset in Jaffa—where the Mediterranean ends—to the Kibbutz Beit Alfa, where I discovered a branch of my family that took a different path ninety years ago, when my grandmother left Grodno heading west toward the River Plate and her sister headed east toward Palestine. Over those five days I came to understand the mixture of identities that my friend Jorge Drexler so beautifully sums up when he says: "I am a Jewish Moor living among Christians, I don't know who my God is, nor who my brothers are."

The border of Jerusalem

My history of mixed and fragmented identities pales in comparison to the human drama that has endured in Jerusalem for millennia, forged in irreconcilable beliefs and passions. This city is one of the great challenges for the Montaignes, including Amit Goldenberg and Eran Halperin from the Laboratory of Psychology of Conflict and Intergroup Reconciliation.

Their experiment closely resembled ours in Vancouver, with one fundamental difference: the discussion was real, not hypothetical, on matters of vital importance to those who were talking. As a result, they were even more careful—if that's possible—to provide the right conditions for good conversation. Before the debate began, all the groups were informed that the others were willing to change their minds.

---

9 Jan Taminiau, fashion designer to royalty and stars like Lady Gaga, said that when deciding how to dress someone he asked them these three questions: Who do you want to be? Who can you be? Who will they let you be? It is an existential view of tailoring and fashion, another good angle on the human condition.

Indeed, a major obstacle to conflict resolution is the assumption that one's *adversary* is unwilling to rectify his position. The result of this belief is that the conversations become more and more rigid: here again we see the self-fulfilling prophecy of reflexivity. Intervening to promote the idea that the other party is persuadable substantially improves intergroup attitudes and the willingness to negotiate, make concessions, and reach consensus. Goldenberg and Halperin took these ideas to their ultimate test, at the heart of the Israeli–Palestinian conflict.

That rift is so deep that merely putting groups in contact for meetings is ineffectual, and sometimes even counterproductive. Those responsible for the experiment searched for volunteer facilitators who were convinced that it was possible to find a solution. Before the meeting, these facilitators were in charge of communicating to the members of the groups that their *opponents* had more flexible opinions than they were imagining. Those volunteers were the Montaignes, the catalysts of consensus.

Goldenberg and Halperin obtained encouraging results, even more considering that the study took place in late 2014, a time of particular hostility for the conflict. They showed that conversation worked when participants were informed that their interlocutor was willing to change their minds. Only then was the group interaction constructive and respectful, when the participants let go of some of their prejudices, and were more willing to cooperate and seek joint solutions.

The variables observed in this study are different from those in our consensus experiment. The two studies were also independent, conducted thousands of kilometers apart, and in very different contexts. However, the conclusions are very similar: those who receive the message that change is possible show more positive attitudes towards each other and this leads to a greater willingness for dialogue, to a softening of their rigid views, to finding consensus, and to accepting serious obligations for the sake of peace.

Mental obstacles

Believing that we are malleable produces a greater predisposition to change. That is the inflationary bubble of reason, further proof of our reflexivity. The opposite process also occurs and, in fact, is much more common: believing that our adversary is rigid is the fastest route to stalled communication. This principle not only applies to our view of people and their ideas: it is also pertinent when it comes to our achievements, our virtues, and our emotions.

The importance of a flexible mind has been proven in a number of studies carried out in schools. Carol Dweck, a professor of social psychology at Stanford, investigated the responses of a group of ten-year-olds to a problem that was beyond their abilities. The children reacted very differently: some understood that the problem was very difficult, perhaps too much for their capabilities at the time, but they were enthusiastic and faced up to the challenge. Those students are the ones with a malleable mentality, also called a *growth mentality*. On the other extreme are those who have an inflexible mentality; they become frustrated and blocked when they encounter a difficulty.

The first group know that if they aren't able to solve a problem in the moment it is presented to them, they can learn what they need to in order to be able to solve it later. The second group feel stuck, thinking if they can't solve it right away they never will. From this perspective, mental abilities are rigid: the impossible today will forever remain impossible.

Dweck followed the educational trajectory of both groups and found that those with a growth mindset usually acquire the necessary knowledge and end up solving the problem. The others, on the other hand, avoid it. They get angry, they cheat, they call themselves incompetent, they get frustrated, and eventually they quit school. The reflexive bubble increasingly accentuates their initial differences, which, in many cases, are ones merely of predisposition, not knowledge.

So does your ability to make progress depend on the luck of the draw in terms of your disposition? The good news is that the answer is no: while some people are predisposed not to, everyone can learn. And sometimes, as in Dweck's experiments, without making an exceptional effort. You can substantially improve a student's educational trajectory by simply showing them that it's possible, and worth the effort, to spend some time struggling to get to places that seem far beyond one's reach. It is the specular version of malleability, applied to oneself rather than to others.

This rule applies both in and out of school, throughout our lives. It even, to a certain extent, applies to other species. Let's start with an example of the latter. The story begins in the mid-twentieth century at Johns Hopkins, when Curt Richter did a somewhat gruesome experiment. He put rats into buckets of water and let them swim there until they gave up and drowned. Most would quit struggling after two or three minutes. Some rats, on the other hand, swam for days before giving up. The difference was that marked: a few minutes versus a few days.

Richter had an idea to explain this phenomenon: hope. The belief that sooner or later a way out of the bucket would appear fueled the motivation to keep swimming. He concluded that this belief was what distinguished rats that swam for days from those that gave up after a few minutes. He then devised a new experiment in which he infused "hope" (or a growth mindset) in rats. He would place them in the water and rescue them just before they drowned. With this simple action, showing them that an escape route existed, he managed to get the rats to swim for much longer when they entered the bucket a second time. Knowing that things can improve alters a rat's behavior. So we can see that this is more than merely a human trait, it is a fundamental aspect of life. In Art Spiegelman's graphic novel, Jews are depicted as mice and they have a wide variety of fates in the concentration camps. Viktor Frankl, in *Man's Search for Meaning,* states that the principle is the same: the key to survival was to find meaning, not abandon hope, and to keep clinging to life even in the most desperate situations.

The writer David Epstein has studied this phenomenon in the sports world. Why are we capable of running so much faster today than we were a hundred years ago? Part of the reason is found in technological advancements, improvements in diet, shoes, and training techniques. But these arguments are not enough to explain why, after someone passes a seemingly insurmountable milestone, others soon appear, as if they've been waiting in the wings, who are able to repeat the accomplishment. What changes is the mentality, the hope factor. At the limit of human capability, knowing that something *can* be done is the last key needed to achieve it. Epstein tells the story of how, for a long time, no one could run a mile in less than four minutes. In fact, until 1950, doctors and scientists believed it was physically impossible, that the human body was unable to survive such an effort.

That idea was firmly entrenched until, in 1954, Roger Bannister ran a mile in 3:59:40. The story, according to Bannister himself, is an ode to the power of the growth mindset: "There was no logic in my mind that if you can run a mile in 4:1.25, you can't run it in 3:59 . . . I knew enough medicine and physiology to know it wasn't a physical barrier, but I think it had become a psychological barrier." In 2021, the world record is held by the Moroccan Hicham El Guerrouj, with a time of 3:43:13, and 1,400 people have run a mile in under the "impossible" barrier of four minutes. Once people are convinced that the barrier isn't unbreakable, they overcome it.

The barrier, of course, was in the brain all along; more specifically, in a brain control device whose tasks include managing our physical resources so they do not exceed a healthy limit, so they aren't reduced to the point where we can injure ourselves or run out of energy. This system is sometimes overcautious. Knowing how this "switch" works allows us to learn how to regulate it in order to surpass the limits imposed by the brain itself. Epstein uses the example of long-distance sport—marathons, triathlons, big climbs—and shows us that the body is more prepared to face this type of trial than we perceive. And that in some cases it is possible to deactivate that part of our brain and achieve results beyond our wildest dreams.

In other realms, many people have experienced situations where they are able to surpass this limit. Sometimes, our bodies will scream that we shouldn't do something, even when there is no real risk involved. These are mirages that the body falls prey to: flying in a plane, riding a roller coaster, watching a horror movie . . . In each of these situations, our brain's regulator tells us to abort because it perceives a danger that, rationally, we know does not exist.

I had one of those experiences on the television program *El cerebro y yo* with Diego Golombek, my co-host. We'd traveled to the province of Salta, where I was scheduled to jump off a bridge in order to experience first-hand how time passes when you are speeding into the abyss. I didn't know, of course, that the experiment would turn out to be quite different. When I arrived at the bridge, equipped with the harness and elastic band that would prevent me from crashing into the ground, I realized there was no way I was going to jump. The feeling of vertigo was overwhelming, intolerable. My entire body was telling me it was suicide and, in a way, my body was right. Throughout the ages, falling from that height meant dying (and still does without the harness). I repeated, over and over again, that it was safe, that thousands of people had survived—and would survive—the experience. It was a long, arduous inner dialogue. A battle between my brain's regulator that had set off all the alarms to convince me not to jump, and the voice of reason (plus my guilt over having an entire production crew waiting to film it). Finally, I jumped. The fall itself was shorter and infinitely less interesting than the raging battle that had taken place in my mind. A malleable mind is a superpower that leads us to do the seemingly impossible, from relatively insignificant things (like my jump from the bridge) to the most extraordinary achievements.

Exercise: Ideas for living better

1. **Explaining your motives helps you make better decisions**
   Tell others why you did what you did. Or, better yet, what you plan to do. By explaining your arguments, it will be easier for you to locate inconsistencies, find alternative ideas, and create more suitable moods for decision-making.
2. **Learn how to converse with yourself**
   When programmers can't find a solution they sometimes resort to the "rubber ducky technique": they explain what they are trying to do to a (usually imaginary) toy, which has no context or computer knowledge. That process of making every aspect of the problem explicit helps you identify important issues you've overlooked, see new perspectives, and find other solutions.
3. **Write down your decisions**
   The mere fact of writing down decisions opens up space for a more deliberated and profound—a less automatic—inner reflection, which helps us to make better decisions. It is another, very effective, way of sparking an inner conversation.
4. **Create spaces to facilitate good conversations**
   Without areas where we can offer and receive criticism, where we can reveal and learn about alternative perspectives, we have fewer options to understand problems fully and make the best decisions. Open the door to those encounters—remember, always in small groups—and foster and participate in them.

5. **Get comfortable with nuance**
   Even in those issues where we believe we have categorical, inflexible opinions, it is likely that, if we look closer, we can find subtleties from which to build bridges. That creates a foundation for cultivating relationships.
6. **Look for the Montaignes, the confident grays**
   People who are able to recognize good arguments on both sides of a debate are helpful for reaching consensus, and for using that consensus to resolve problems. Whatever your stance, Montaignes are always great allies because they help conversations flow.
7. **Assume that everyone can change their mind**
   Seeing others (or yourself) as incapable of changing their point of view is a dead end. It degrades conversations (even inner ones) and makes it difficult to move toward a solution. Recall a case where your opinion has changed over the years and admit that it could happen again. In the proper context, in the face of good arguments, others can change their minds as well.
8. **Get used to not knowing everything**
   Expose yourself, every once in a while, to unfamiliar situations or new problems. It will train you to get accustomed to the altitude, like mountain climbers who make intermediate ascents before attempting a big climb.
9. **Remember that some limits are mental**
   Some things have never been done because they aren't possible. Many others have never been done because the fact that they've never been done creates a mental barrier that seems insurmountable. Prepare well and take a risk: give it a try.

CHAPTER 3

Our Life Stories

How to edit our memory and discover who we are

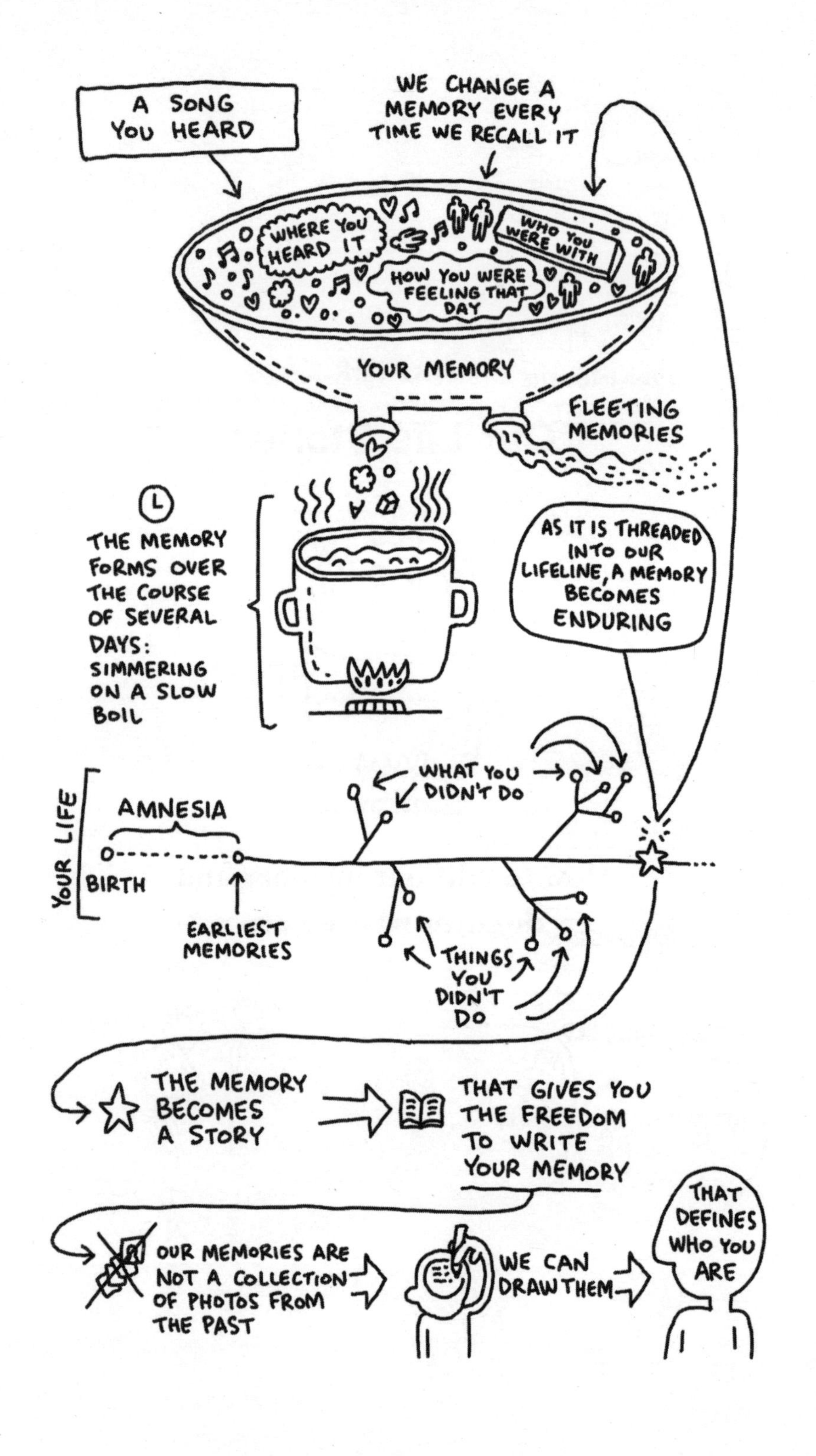
A SONG YOU HEARD
WE CHANGE A MEMORY EVERY TIME WE RECALL IT
WHERE YOU HEARD IT
WHO YOU WERE WITH
HOW YOU WERE FEELING THAT DAY
YOUR MEMORY
FLEETING MEMORIES
L
THE MEMORY FORMS OVER THE COURSE OF SEVERAL DAYS: SIMMERING ON A SLOW BOIL
AS IT IS THREADED INTO OUR LIFELINE, A MEMORY BECOMES ENDURING
YOUR LIFE
AMNESIA
BIRTH
EARLIEST MEMORIES
WHAT YOU DIDN'T DO
THINGS YOU DIDN'T DO
THE MEMORY BECOMES A STORY
THAT GIVES YOU THE FREEDOM TO WRITE YOUR MEMORY
OUR MEMORIES ARE NOT A COLLECTION OF PHOTOS FROM THE PAST
WE CAN DRAW THEM
THAT DEFINES WHO YOU ARE

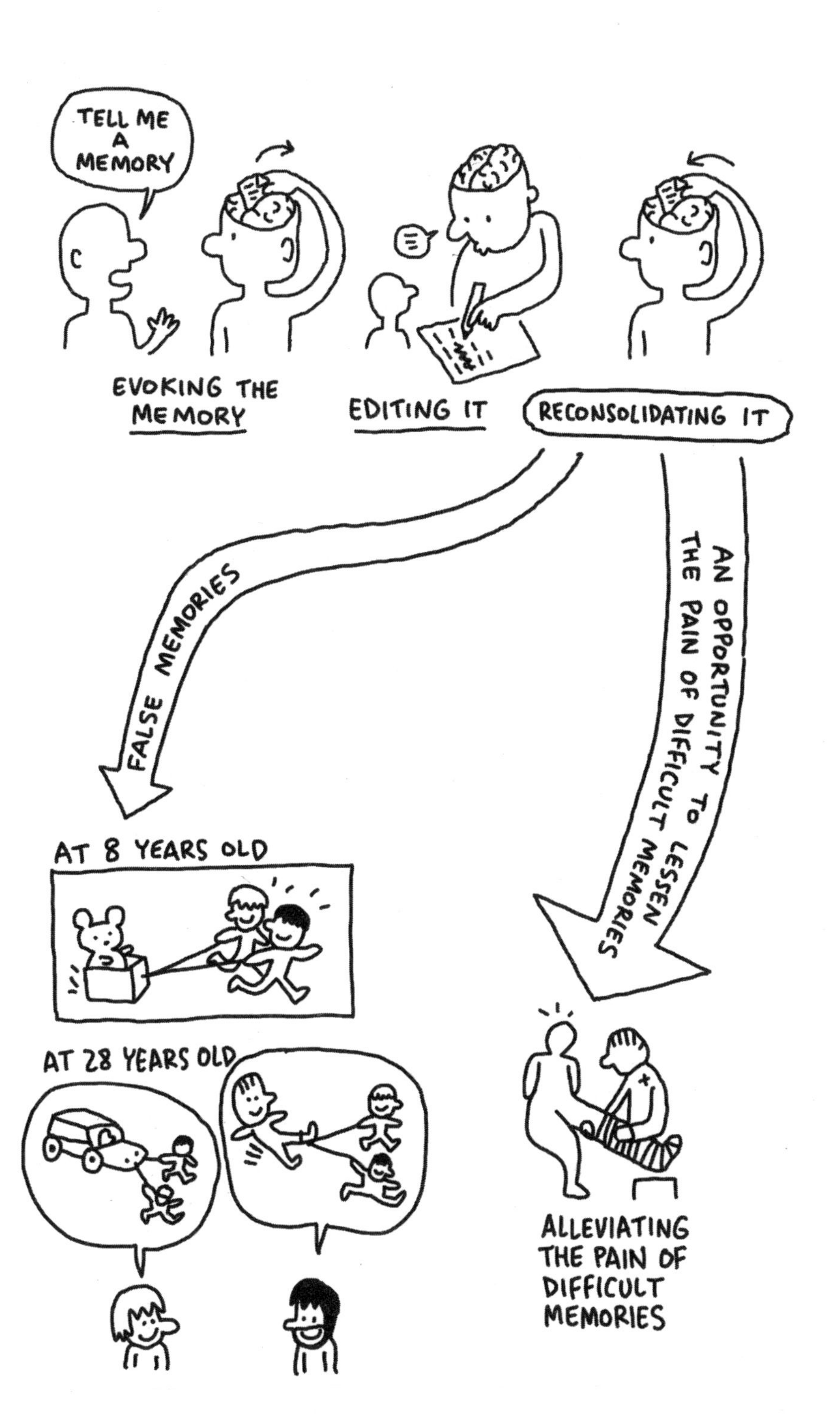

TELL ME A MEMORY
EVOKING THE MEMORY
EDITING IT
RECONSOLIDATING IT
FALSE MEMORIES
AN OPPORTUNITY TO LESSEN THE PAIN OF DIFFICULT MEMORIES
AT 8 YEARS OLD
AT 28 YEARS OLD
ALLEVIATING THE PAIN OF DIFFICULT MEMORIES

Our memory is capricious. We remember things we'd prefer to forget and we forget things we'd like to remember. Some memories grow and become beliefs. A teenage girl who feels she's being made fun of at a party may "construct" the memory, out of that biased perception, that something is wrong with her body or the way she speaks or dances. And that belief leads her to marginalize herself, based on a stigma that is sometimes created by others and that, often, we create ourselves. It seems like memories are automatically recorded, but that's not how they work. We have a certain freedom of choice when inscribing each of those episodes in our memory. Conversations, particularly the ones we have with ourselves, are also a fundamental tool for editing our memories and defining our identities. Our memories aren't merely a collection of photos from the past; surprising as it may seem, we can actually draw them. Our memory is filled with fiction, connections between distant points, and interpretations.

We will see that memory is, above all, a creative exercise of linking life's episodes into a continuous story. Of weaving each memory into a plot that forges our identity. We remember a song—or a place, or a story—not so much in its own isolated capsule, but by how we relate to it: where we heard it and who we were with, what was happening, and how we felt that day. That is why before developing a notion of "oneself" there is only amnesia. We remember nothing of those decisive first years of our lives because the memories had no identity to anchor themselves to. They were memories that belonged to no one.

In the assembling of memory, conversation again plays a fundamental role because the story of who we are, what we want, what we can and cannot be, only takes on true meaning when we explain those things to other people. At some point in our childhoods the propensity to share our experiences spontaneously appears. We talk about some things and keep others to ourselves: through this selection process, each of us begins to mold our persona; a story, a saga, is created. It is not the story of Harry Potter or Tintin or Sherlock, but our stories of ourselves. These stories are identity centers where we each organize our memories.

We are all writers and editors, and we configure our identities over time in sequences of conversations with ourselves. In those stories we blend together, unconsciously, fiction and reality. That is how false memories can appear: they are things we remember with enormous conviction, but which never happened. We will see that they are built through a very precise mechanism: every time we evoke a memory, our memory bank becomes labile and can be rewritten. False memories are the result of a sophisticated and creative system that can allow us to somewhat freely draw the outlines of our identities—a system that makes memories less precise but at the same time more malleable.

Minos, King of Crete, enraged by the murder of his son, condemned the Athenians to send fourteen boys and girls into the Minotaur's labyrinth every nine years. After several voyages, Theseus, Prince of Athens, decided to join a group of the sacrificial children and put an end to the Minotaur. After killing it, he was able to escape with the help of Princess Ariadne's ball of string and return in his ship to Athens.

After the festivities, the Athenians honored the vow they'd made to Apollo: to travel annually to the sanctuary of Delos in Theseus's ship. Over time, some of the ship's planks decayed and the Athenians replaced them with new, stronger wood. Years later the ship no longer retains a single one of its original planks. Thus arises the logical question: is it possible still to honor the vow? Is the ship still the ship of Theseus? And if it isn't, at what exact point did it cease to be? Is it true, as Heraclitus said, that a man cannot bathe in the same river twice because the second time he is no longer the same man nor it the same river? And if they were to construct another ship with the original planks from the first one, which of the two would be Theseus's ship? These questions are the basis of our outline of the blurry limits between identity and memory. How is it possible for them to persist on a foundation of deteriorating, shifting sand?

Creativity begins in our memory

In his chronicles of the Gray Angel, Alejandro Dolina tells how the mediocre angel one day warned a pharmacist that he would die on a Friday. The pharmacist happily accepted the prediction: on his "days of immortality" he took all manner of risks and on Fridays he was extremely cautious. But what happened was this: "On Thursdays he would visit his friends and relatives to say goodbye. On Fridays he would go mad and scream, begging for mercy. On Saturdays he would get drunk to celebrate his luck. Things went downhill quickly. Herrera had to close his pharmacy, fell into poverty and acquired a deserved reputation as a crackpot. He committed suicide on a Tuesday, to the satisfaction of those who uphold the doctrine of free will. The Legend Refuters attempted to use this story to prove the nonexistence of the Gray Angel, although it serves more as proof of his inefficiency." This humorous anecdote illustrates a typical mistake: refuting a principle based on a one-time failure. And in this same way, the tedium caused by rote teaching has led many to conclude that educators should, pun intended, just forget about it. In fact, memory has long ago lost its prestige; it lost the *battle* against creativity in the world of education. But can creativity exist without memory, or memory without creativity? I will argue here that the answer is no, by introducing a different view of memory. Not so fixed, closer to painting, drawing, poetry. Instead of drawing on a canvas, we all paint in our own brains.

The Muses and memory

Linking memory and creativity is not a new idea, but we often forget it. In some examples we see how this idea traveled from the origins of our culture to the present day.

In Greek mythology memory was personified in Mnemosyne, the daughter of Gaia, the Earth. This makes sense, because the very foundation of existence is rooted in memory. Mnemosyne is the mother of the Muses, those deities that the Greeks and Romans turned to for inspiration. The eldest of the Muses is Calliope, the goddess of epic poetry who gives rise to all narrative. In other words, in Greek mythology creativity is born from memory and memory, in turn, from the Earth.

This story that interweaves memory and creativity also hints at their separation, which persists to our time. The Muses are external figures of memory, a sort of Wikipedia. The *Odyssey*, in fact, begins with the lines: "Sing in me, Muse, and through me tell the story . . ." Later Plato referred to the figure of the poet Ion as a piece of inert iron that is magnetized by contact with the Muse and thus able to attract words and, with them, people's souls.

The idea underlying these metaphors is that creation and inspiration do not emanate from our own memory but rather require another substance. This would be a mere historical curiosity if not for the fact that these ideas are still vividly alive in our way of thinking and feeling about creativity. The image of a person waiting patiently on a riverbank for inspiration to arrive, for an idea to appear, as if touched by a Muse or whispered to by an angel, is a familiar one to us. Thus we see how we've inherited this idea.

Paul McCartney has often said that the melody for "Yesterday," one of the twentieth century's most magnetic compositions, came to him in a dream.[10] Even when his ideas appear suddenly in the

10 This is not McCartney's only dream-inspired song. "Let It Be" begins with the famous lines: "When I find myself in times of trouble, Mother Mary comes to me, speaking words of wisdom: Let it be," which were also found in an oneiric encounter; in this case with Paul's mother, Mary, who died when he was just a teenager. When he composed "Let it be," McCartney was suffering over his latent conflicts with John Lennon and his mother appeared to him in a dream to reassure him that everything would be OK.

middle of the night, McCartney isn't confused about where they come from. He considers that dream the expression of a memory. The Muse was inside him, in his memory. That was why he went out to the record stores of Liverpool, to see if he could find that melody, which he suspected he had heard before. But it turned out that it didn't exist: it had been, in effect, an oneiric creation built of fragments and ideas floating in his memory.

Even more evocative is the story of "My Sweet Lord," the first song by George Harrison after the breakup of The Beatles, which immediately became a huge hit. The problem is that the song was highly similar to another published in 1963 by The Chiffons titled

"He's So Fine." So much so that the case was brought to court and the judge ruled that "in musical terms, the two songs are virtually identical." Harrison knew the other song, but he hadn't copied it intentionally. He was charged with *subconscious plagiarism*. This famous case acknowledges the confusion in the creative process: the melody emanated from Harrison's memory, but he felt it as a result of creative improvisation, the touch of the Muse.

Creativity, the final delusion

The moment we are alone we begin to create voices, sometimes very simple ones, that talk to us about what we've done, what we have to do, or what we wish we'd done differently. It would seem that recognizing ourselves as the creators of those voices is simple, but it isn't. We need look no further than how we "forget" that our dreams are our own creations, and that is why we experience them as so different.

Julian Jaynes, professor of psychology at Princeton, came up with one of the most provocative theories of the philosophy of the mind based on this idea. His argument is that when Homer introduces himself as an instrument of the Muses, or when his characters listen and obey the gods' voices, he is not using metaphorical language. Jaynes observed that this description was repeated throughout the cultures of that period and as such reveals a fundamentally different way of thinking. When Hector obeys Apollo's orders, whispered to him through different characters, what we are seeing—according to Jaynes—is the functioning of a schizophrenic mind. After an exhaustive analysis of the ancient texts—our fossil records of human thought—Jaynes suggests that our ancestors lived in a garden of schizophrenics.

Before Homer, people did not recognize their own voices and ideas as their creations. That stage is what we call *primal consciousness* and we now understand it as a characteristic of schizophrenia or dreams. Then, about three thousand years ago, in the Axial Age, a profound social transformation took place in India, China, and the West. The German philosopher Karl Jaspers described this era, which saw the emergence of the philosophies, religions, and societies that are the foundations of modern culture, as the most abrupt transition in history.

The engine of such change was the invention of writing, which gave memory a rigid support and kept the stories from degrading when passing from word of mouth, thus stabilizing thoughts. It was an early Wikipedia. The written medium made it possible for people to begin to *see* their own voice on paper or etched in stone, making it its own entity. This led to the emergence of consciousness as we understand it today: we are the authors of the voices in our mind. But the old way of thinking left a mark that is still visible: creativity. When a surprising idea *appears*, we continue to perceive it as sparked by external inspiration. This residue of Homeric thought confuses us, leading us to forget that our factory of ideas, its raw material, resides in a fabulous pinkish-gray mass sheltered in our skulls.

First times

Ideas always emanate from the brain. They are the result of information that we have incorporated throughout the course of our lives. Sometimes an external event triggers an idea. However, we often confuse the message with the messenger: those external stimuli are not the source of the idea, they merely enable us to search for it in our memory.

We can think of memory as a huge, almost infinite, table covered with Lego pieces of all sizes, shapes, and colors. The creative act consists of efficiently scanning this table to find pieces that combine in an attractive way. That's essentially what happened in Paul McCartney's dream. Each of the elements of "Yesterday" were already in his memory bank, in chords and melodies he had heard. The dream merely turned out to be a good context for combining pieces that he hadn't been able to put together in such a successful way in his waking life.

Creative ideas spring from a process of searching through the labyrinth of memory. So, how do we find a good idea that's capable of appealing to us and to others from an infinite number of possible stories? When we transfer the problem of creativity to a problem of memory a simple recipe appears to make a complex task easier:

reduce the search area so we don't get lost and wander indefinitely in the abyss.

Very general questions result in creative paralysis. For example, when you ask a child, "How was your day?"—or when you're asked that, depending on your perspective—the response will be, at best, monosyllabic. And the same thing would happen at a dinner party if we surprised a guest by asking him to tell a story. The person would suddenly find himself between a rock and a hard place, unable to come up with any story, perhaps waiting for a Muse to appear and lend him a helping hand. In the end, chances are he won't come up with anything. Not because the Muse hasn't shown up, but because vague questions are a terrible way to activate your memory, where all our stories live, just in no particular order.

The results change radically if we search for a story in a specific corner of our memory. A good example of this is stories about our first times: our first kiss, our first trip abroad, our first love. If you try this exercise at a table shared with people you don't know very

well,[11] you will see that, all of a sudden, as if galvanized by Calliope, everyone at the table becomes a great storyteller. Everyone manages to connect with the stories inside them, with an amazing story about how they prepared for or were surprised by their first kiss; what brought them to that moment, what happened next, and how they felt. The story tends spontaneously to have an effective narrative structure. Jacobo Bergareche wrote a book around this idea: *Estaciones de regreso*. The narrator travels back to concrete points of his life's timeline, allowing well-marked entrances into his memory, the natural starting place for storytelling. The story we create about ourselves rises to the surface of our consciousness before again plunging back into that blurry and contiguous plot of most of the rest of our days.

Another example of how a specific prompt triggers memory and, with it, creativity is seen in an experiment where we asked one person

11 A wedding is the perfect place to try this experiment.

to tell another a memorable story. We secretly told some listeners to follow the story with rapt attention while telling others to ignore it. They could look at their phone, get distracted by other conversations, interrupt with irrelevant questions; basically we were asking them to do what we all do often, at work, at home, or with friends: totally disregard someone who is speaking to us, despite their passion.

We found that the experience of the storyteller varies greatly depending on the attitude of the listener. It changes everything, even their own judgment about the story they've shared. That is why the "like" economy is such a hellish trap. We don't judge what we've written or told someone on its intrinsic merit, but on the intensity of the applause we receive for it. The most striking aspect of this experiment was not so much the result, which was predictable, but the scale. The exact same story is either a sublime success or a resounding failure merely according to the attention it receives from the listener. This is another creativity killer. Stage fright is the main reason we abandon our childhood creative expressions: drawing, singing, playing, dancing.

Just as we saw with "first times," guidelines here were also essential to drawing out good stories. With simple prompts such as: "Tell us about something that recently embarrassed you," participants came up with extraordinary, flowing, sometimes emotional, stories; without the prompt, they displayed total paralysis. We witness this in conversations with friends we see every day: we can talk for hours; we never run out of spontaneously appearing topics to discuss. On the other hand, when we meet up with someone we haven't seen in a long time, with whom we need to catch up on every aspect of our lives, we don't even know where to start. So we end up talking about something trivial, like the weather.

The axiom of choice

What happens in these conversations is that the drawer filled with possible stories is so large that we are overwhelmed by the choice. We are faced with the false sense of freedom of a blank page, of which Deleuze said: "A canvas is not a white surface, I think all painters know this. The canvas is covered in clichés."

This confusion is not exclusive to the creative process; it is a general problem of thought. When making a decision, excessive options are experienced more as a curse than a blessing. Examples abound: an infinite wine list or menu; a store with millions of colors, models, and prices. It might seem that having more options is always better; after all, we can cast aside the ones that don't interest us. What happens, however, is that most of us are no good at this seemingly simple casting aside. Therefore, having many alternatives doesn't give us more freedom. In fact, it usually paralyzes us. This is the paradox of choice.

The caricature of this dilemma is the French philosopher Jean Buridan's donkey, who ends up dying of starvation because it is unable to choose between two identical bales of hay. If it had been given just one, of course, there would have been no conflict; but when offered two equal options, instead of randomly picking one over the other, it attempts an impossible task: determining which one is better.

We don't die of starvation because the brain ends up solving the stalemate by injecting random ion streams into the circuits that encode each option. But this process is slow and almost always inefficient, so it's a good idea to help the brain accept that the answer is random. My colleague Jérôme Sackur always uses a coin. Every time he is faced with equivalent and unimportant decisions, he tosses a coin, highlighting the stalemate and the inevitable role of chance. This is a classic method to improve our decision-making, so we don't suffer the same fate as Buridan's donkey. To implement it we need only reach into our pockets, and it relieves us of the responsibility of choosing, thereby reducing stress and saving us time.

The paradox of choice is transversal and foundational to thought. In fact, it is the core of a central conflict that lies at the basis of modern logic and mathematics. In 1904, the German mathematician Ernst Zermelo formulated his now famous *axiom of choice*, according to which, given an infinite series of sets, one element can always be chosen from each of them. It doesn't seem like a big deal. For example, you can choose the first element of each set, or the smallest. But it turns out that the sets formed by any of the things we can think of don't have to be ordered in such a way that there is a first or a last element, or one to the left of another. Nor do they have to follow an order by color, size, or price. And in that jumble you can't guarantee a search procedure that will always work, as the set becomes larger and larger.

Zermelo's axiom of choice begins with an intuition and becomes more and more mysterious as we approach its logical implications. Twenty years later, the mathematicians Stefan Banach and Alfred Tarski showed that, if the axiom were valid, then it would be possible to take a sphere, divide it into parts and with each part build two spheres identical to the original. This is perhaps the best evidence of the difficulty of choosing in vast spaces. By taking this issue to the limits of infinity, mathematics shows that the choice is completely impossible. If it weren't, absurd things, like duplicating matter, would happen. That's how strange our ideas and choices are in the—also highly intricate—labyrinths of memory.

Memory as a creative exercise

Experiments in conversation, emotional accounts of our first times, and Paul McCartney's dreams show that memory is the fuel of creativity when it comes to coming up with and telling stories. Now I would like to analyze the opposite relationship: creativity is also the fuel of memory. I would argue that creativity and memory are like yin and yang.

Let's begin with the scenario where this idea is questioned most often: in formal education. Most students lament: "Why do they make me learn the name of every river in Asia? I'm never going to need to know that! I'll probably forget it right after the test and, besides, I could just google it." These days, our collective memory is found on hard drives floating in the cloud (those are the Muses of our time). Their argument seems perfectly logical, but it actually contains a large misconception. Let's see it in action: in the middle of a heated negotiation, we remember having read something that may tilt the odds in our favor. Do we ask our interlocutor to wait a few minutes so we can look it up on the internet? No, because in negotiations, as in friendship, love, and life itself, we can't take time out for research. The perfect word is perfect because it appears at the perfect time. In life there's no pause button, like in *The Matrix*, to momentarily stop reality and download a program that allows us to learn a new language or martial art.

Learning the names at school of all the rivers in Asia isn't important because of what knowing them means; what is important is that students develop the tools to efficiently "draw" a memory so they can recall it at will and without effort. Basically, this is learning how to think, how to connect new knowledge with previous knowledge through a coherent story. This ability gives rise to deep learning, which some modern education researchers consider the antithesis of inert learning, which remains disconnected from all our experiences and prior knowledge, like inaccessible islands floating in lost parts of our memory. Inert knowledge can only be recited back. We can't put it into practice or see it from another perspective. If we review our realms of knowledge (such as mathematics, history, economics, science, and, above all, what we think we know about ourselves) and ask ourselves which we know deeply and which inertly, we see that those realms that we identify with inert knowledge usually coincide with what we don't think of as our natural abilities. This can be changed, although it takes work, of course. The way we would do it is by organizing the knowledge in those realms into stories, looking for logic, geometry, how to connect the dots.

Let's look at simple ideas for achieving this by going back to our example. To learn the names of all the rivers in Asia, it might be useful to shrink the list and make it comprehensible; to understand who lives near each river, how they've changed the history of the regions they run through, how they divided the peoples who settled on their banks, how they connect to other rivers, and what would happen if they dried up or became polluted. If we give them context, meaning, and a story, the elements on lists are more easily remembered. It is also important to remember that properly learning the Asian rivers, or the periodic table, or the phases of the industrial revolution, or the structure of parliamentary governments should be an exercise in learning the logic of memory, just as when we play sport as children it's not in order to become professional athletes, but to train a series of faculties associated with sport such as physical fitness, speed and endurance, motor coordination, spatial sense, strategy, and teamwork.

The problem is not the teaching of memorization, but the way it is usually taught. We are well advised not to be misled by this confusion and give way to the lazy temptation to abandon our attempts to improve our memory. The reality is that there are few more pertinent exercises than teaching us how to build those stations from which we can summon and describe knowledge. Once those founding stations have been formed, each of us will be truly free to choose what to fill our memory with.

That is my premise. All freedom is built with tools. The freedom to express oneself requires the proper use of language; in order to devote ourselves to painting, we have to perfect our hand movements. In much the same way, very high on the list of instruments that allow us freedom to think are those that permit us to write and read our memory.

The geometry of memory

My central argument is that learning to write one's memory is first and foremost a creative exercise. The best way to explain this phenomenon is to see how those with prodigious memories shape their recollections. Our natural response is to imagine that they have enlarged their memory bank. The larger the bank size, the more memories they can hold in it. But the science of memory—which has accomplished a lot—shows that it doesn't work that way. People with good memories do not have larger memory banks; rather they find ingenious and creative ways of storing their memories and creating clear paths to reach them. Having a good imagination makes it possible to establish associations, relationships, and bridges that allow us to establish more effective memory.

Like Mnemosyne herself, this idea dates back to the ancient Greeks and the Greek poet Simonides of Ceos's "memory palace." The story goes that the poet left a banquet to contemplate the sky and, at that moment, the palace collapsed. In the midst of the ruins, asked to identify the victims, Simonides was amazed to find he remembered in great detail where each guest had been sitting. He could perfectly reconstruct that image. The paradox that Simonides discovered is this: remembering a list of names is impossible, while remembering something that entails more information—in addition to each of the guests, the precise place they occupy at the table—is much easier. From that experience he came up with a general rule: space is the natural terrain of memory.

This is true for people of every culture around the globe; for children and adults, and also for a large number of species of the animal kingdom. Until the invention of the word—a blip in the history of life—space was the grid of memory. The idea of the memory palace uses this natural framework to organize any memory, however abstract it may be, even if it seems to have no relationship to the space itself. This operation requires imaginative architectural work, like that carried out by the dream engineers in Christopher Nolan's film *Inception.*

The first step is to build a mental palace with different rooms that can be easily identified and mentally navigated; many people choose their own home. This will be the repository for any list of words. Suppose, for example, that you want to remember *giraffe, raspberry, chisel, dynamite, bolero, tiramisu, calculator . . .* Simonides's technique consists of "inserting" each item on the list into a room of the palace. Imagine a giraffe in the first of the rooms; the entryway, say. The more impactful, emotional, vivid, and even scatological the image, the better the tool works. A twisted giraffe stuffed into that tiny space, forcing us to make our way through the narrow pocket

of air between the wall and its butt. Now that's a powerful image that will stay with us. We continue walking through the palace and to the right, in the second room—a bathroom, say—there is a thick liquid dripping off the mirror. It looks like blood until we lick it and it's unmistakably raspberry sauce. So now these unconnected elements on the list are suddenly organized within the architecture of the space. By practicing this technique, anyone can greatly improve the number of words they are able to remember.

It's like making a cubist collage with intense images to carve a list of names into our brains. Memory is not built by repeated efforts, reciting words over and over again *ad infinitum*, to furrow deep grooves in the brain: memory is built, essentially, with creativity.

The memory palace is not the only way to organize unrelated words, memories, or ideas. In general, the technique for organizing memory consists of uniting a series of disparate elements within a story. This is common to all mnemonic rules. For example, the four bases of the genetic code are tabulated with the letters A, T, C, and G. Without any clear relationship, these letters are often confused. To remember them better, people in Buenos Aires associate them with the initials of Aníbal Troilo and Carlos Gardel, two great figures in the world of tango music. Of course, each culture finds its own way to turn these letters into a story. This is a good exercise using creativity to aid memory. To be more precise, the good exercise is not applying this rule, but learning to replicate it by finding good examples for anything we need to remember.

We can "visualize" this idea in the following illustration, thinking of the two objects in the upper panels as lists, each segment representing an element we want to remember. In the lower panels, the "lists" are mixed together with all our other accumulated information. When these objects are set among the background noise, a significant, almost magical, difference occurs. We see the one on the right automatically, effortlessly; it would be hard to miss. Yet, in order to detect the one on the left, we have to make a serious effort and, as soon as we identify any of its parts, the rest vanishes. The logic of memory is not very different from the logic of perception. The object on the left illustrates inert knowledge, the things we struggle to commit to memory, that have no relationship whatsoever to one another. The list on the right, however, has each segment naturally and geometrically linked to the next. The art of memory consists of ordering and giving coherence to fragments of knowledge, either in a story or an image: it is about shifting, moving, and rotating the pieces we want to assimilate so they acquire logical continuity and can be remembered automatically despite the background noise. Anyone who understands the periodic table, the logarithmic formula, the events of the French Revolution, or the grammar of language, understands them in this way. As if the facts leapt out at them from their memory bank.

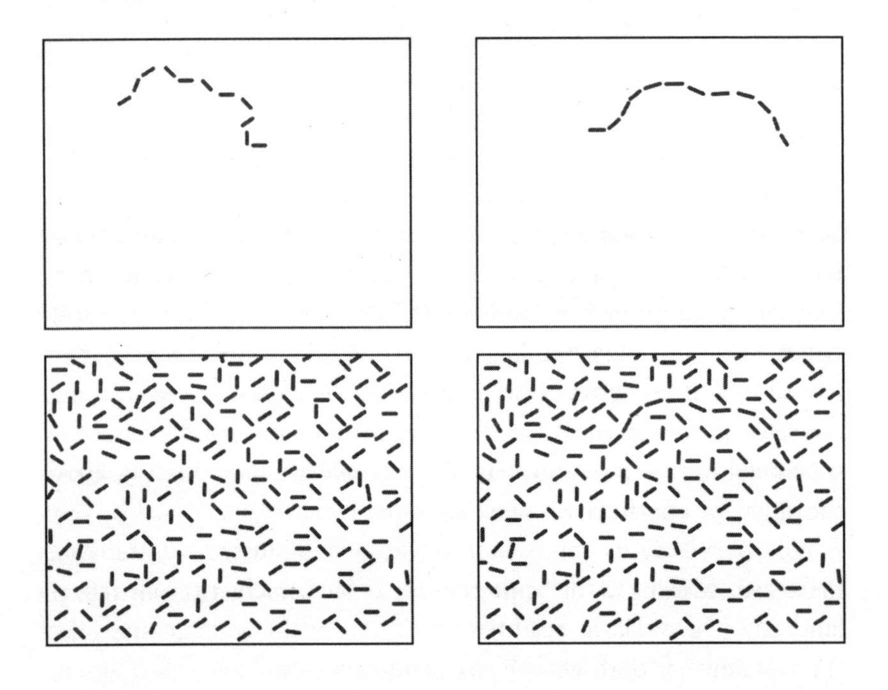

These are cartoonish generalizations about the art of memory. Normally, we connect acquired knowledge in a much more complex way. But the underlying idea is the same: we have to build a meaningful story in order to link new knowledge to the things we already know. As our approach to memory becomes more creative, we deepen our capacity for learning.

Our life timelines form the basis to which our experiences and knowledge are progressively added, like branches off the tree of memory. The framework of this tree is drawn with strokes of fiction that give it coherence. That is why memories are filled with illusions. As I said, the art of memory is not that different from the art of painting. Knowing this offers us a great opportunity to shape the stories we tell ourselves. The same fact can be integrated into very different stories that change our experiences of them and the way we view the future. It is about *acquiring a deep knowledge of ourselves* and becoming the architects of our own memory.

The origin of words

In recent years freestyle rapping has become fashionable, a variant in the very long tradition of improvisation. These disciplines remind us that every conversation is a fabulous exercise in improvisation. With very few exceptions, we aren't reciting some memorized speech when we talk. We improvise constantly, looking for relationships and connections in the memory palace. And all this happens at lightning speed. Each word is produced in a fraction of a second, and while it is being articulated, the following ones are already being conceived. Perhaps most surprisingly, the whole process happens as if nothing in the brain is working to make it happen. We don't feel the slightest effort. That's why we aren't constantly amazed by that mundane yet highly sophisticated form of improvisation we've all mastered, making it difficult for us to appreciate the *miracle* of language. Our task, as scientists, is to dispel the miracle and keep the wonder. For that, we use examples like this:

1. *There are people who prefer cholocate to beer.*
2. *You misread at least one the words in 1.*
3. *You misread again in 2.*
4. *Yes, you read it wrong. Take a good look.*
6. *I saw you smile.*
7. *But you didn't see that we skipped 5.*

8. *You misread again in 7.*
9. *Just pulling your leg, you read 7 right.*
10. *On a scale of one to ten, how well do you think you read?*

The gears of memory and language are visible when they fail, as in the previous example, or in the stuttering of a child, or in the words that get stuck "on the tips of our tongues," a phenomenon to which Roger Brown and David McNeill devoted an extensive study titled, without beating around the bush, "The 'Tip-of-the-Tongue' Phenomenon." Their pioneering work, carried out at Harvard in 1965, established an objective way of measuring a phenomenon that we all experience and about which we have unfounded gut feelings.

In their experiment, Brown and McNeill laid out a bunch of rare, infrequently used definitions and asked a number of people to point out which words they described. Almost 15 percent of the time the participants had a clear feeling they knew the word but couldn't name it. They knew some of its letters and even how many syllables it had and which was stressed—as if they retained a blurry memory of its phonological outline—but they couldn't come up with the word.

The behavior of this phenomenon raises thousands of recurring questions: Why do we forget some names? How does this change as we age? Does it happen in every language? Does music help jog our memory? Sometimes it seems that the key is to make no effort to search for the word we want to remember, so should we just wait for it to show up on its own? Do psychoactive drugs make the problem worse or actually help? Do words on the tip of the tongue have anything to do with Alzheimer's? There is even a study where Julia Simner and Jamie Ward show that those who, through synesthesia, perceive a taste when they think of a word will taste it even when it is only on the tip of their tongues.

Skipping past nearly forty years of research, I'll go straight to a study by psychologists Trevor Harley and Helen Bown in which they asked why some words are more likely than others to get stuck on the tips of our tongues. Among a vast set of definitions, they found that the ones that most frequently inhabit that limbo are those used infrequently, both in the language we produce and in the language

we hear. They are also usually words that are long or difficult to pronounce.

My grandmother—who, by the way, at a hundred and two years old[12] has an extraordinary memory—will go through the entire list of her grandchildren's names before she gets it right. If she just chose a name at random, she would reach the correct one sooner. It seems that the most difficult words to recall are those camouflaged by similar terms, like when we confuse the names of our nieces and nephews, our children, or our grandchildren. However, confusing words is not the same as having them on the tip of your tongue. It is

12 She was a hundred and one when I first wrote this sentence; she'll be a hundred and three when this book is published.

not that my grandmother didn't know my name, but that other names from the list she had saved in some corner of her brain managed to jump the line. Harley and Bown tested out the opposing hypothesis: that the most difficult words to conjure up are those that rhyme less, or are somehow less related by their sound to other words.

They showed that the fragments used to save a word in our memory are the same ones that are then used to retrieve it. We now go to the very origin of that loop, to the first words we remember and say. How do they organize the structure of all our memories?

The arrow of time

To understand the structure of a building, it's helpful to look at photos from the start of its construction. In the same way, observing our infantile thinking is a good way to reveal the mysteries of our minds. The architectural metaphor exposes the difficulty of this. It's easy to program a camera to shoot one frame every fifteen minutes to record the progress of a building. On the other hand, visualizing the development of ideas over the course of a lifetime entails a more sophisticated and careful design.

Argentine photographer Diego Goldberg set out to record images of the "construction" of his life in a project entitled "The Arrow of Time." Over forty years, his family members would meet every June 17 in the same place: they would sit in the same chairs, and compose the same picture before taking the photo. But each image recorded a change in their facial expressions, their bodies, reflecting the traces of the year that had passed.

In the MIT Media Lab, Deb Roy designed a kind of digital panopticon to record the development of his firstborn son in every moment of his life. He filled the house with cameras and microphones to record the little boy's every conversation and every movement: his first step, his first laugh. With that huge pile of hard drives and some programming ingenuity, he was able to identify the precise time and place where each of his son's words was born.

As always when an observatory is established—especially when the subject has no way of expressing consent—the ethics of the experiment are called into question. The ethical dilemma inherent to that situation is complex and its limits are always blurred, since each parenting experience is, in itself, a new venture. I had the incredible luck to share a TED stage with Judit Polgár, the most extraordinary female chess player of all time and one of the most emblematic cases of a child prodigy. Her talk began: "My parents decided I would be a genius before I was born." It is an impressive testimony to the cultural, emotional and family forces that operate at the edges of the human condition, and to the idea that we are all, by default, essentially the result of an unprecedented experiment.

Let's go back to Deb Roy's "panopticon." Many years later, it was thoroughly reviewed by an ethics committee. After all, recording with fixed video cameras is much less awkward than our current omnipresent cell phone cameras, and all of us parents are guilty of pulling them out at times that should be more intimate. Some clear examples are our children's first hugs or their first laughs. These are precisely the moments when we should be most present, not interrupting them by whipping out a camera. Deb Roy instead recorded all of his son's life, all the time, and then edited the most relevant ones, allowing him to keep his full attention on his son. Was that wrong?

From the first day of his life, the microphones recorded an orchestra of vocal expressions that gradually become more sophisticated: cries, laughter, screams, syllables, words, phrases. The first words he utters are, of course, the ones he hears the most. Repetitions stick. And the fact that they are phonetically simple words further contributes to their retention; in a way, this is historical and cultural learning. Languages often have simple words for the objects you learn to name. Simple words have another great virtue: they are always used at precise times and places, helping to anchor them and convert them into the foundations of a memory palace; for example, those that refer to food are only spoken in the kitchen. The opposite is true of prepositions: they are very frequent, but scattered equally throughout both time and space.[13]

---

13 In French we dream *of* birds, in Spanish we dream *with* birds, in Italian we dream birds. However similar their origins may be, the use of prepositions varies enormously depending on the language. To learn them, you have to work at it for a long time. Or during a long time. Or over a long time.

All of these findings are already well known. However, new technology allows us to endlessly pose new questions. Cameras and microphones film the baby growing up, but also his mother, his father, his nanny . . . What are the effects of these other actors on the child who's learning his first words?

Roy's research shows that, while a child is learning a word, the adults around them isolate that word from others that could camouflage it. For example, the word *water* can be used in long sentences such as "Please can you pass me the water that's at the other end of the table?", but when a child is in the learning phase, those around him usually limit their usage to the word itself or, at most, put it with just one other word: "Want water?", "Drink water," or "Hot water."

Which came first, the chicken or the egg? Does a baby learn a word because the adults highlight it in isolation from other words, or is it the adults who learn how to teach? Both answers may be correct, and this is one of many examples of simultaneous learning. And there is a certain creative virtuosity (once again!) when inducting a word in our memory. Protecting it from the degradation produced by contiguous words, understanding that the silence that follows a sound is necessary for it to acquire an identity. Form and content, highlight, underline. All of us learned to speak this way, through our parents' gentle verbal caresses. We will see how each of these words becomes, in turn, the scaffolding of memory.

From here to there

When we moved to Spain, our children were four and six years old. History repeats itself. I moved to Spain when I was four, with my six-year-old brother. There are myriad similarities as well as myriad differences between these two stories that I experienced from such different perspectives. I spent my childhood in Barcelona, my teenage years in Buenos Aires. My twenties were spent in New York and Paris and then I returned to Argentina, to the neighborhood of Villa Crespo, where my children were born, grew up and formed their first words. Until one day we told them that we were moving to Madrid, to a new school, to a new house, to a new life that would begin on the other side of the Atlantic Ocean.

In the months leading up to the move I enthusiastically told them about all the exciting things we would see in Europe. Until my mom reminded me of something I hadn't realized, something that is now in the very essence of what I'm writing. Don't focus on what will be different—whether good or bad—but instead focus on what won't change: that cluster of things that I assumed were obvious but which, from my children's perspective, weren't. For a small child, crossing the ocean is like going to Mars. Tell them, although it may seem obvious, that in the new place there are also trees and cars, elevators and ice cream, buses, doorknobs and hamburgers, laughter, movies, lights, ravioli and parks and games, and books and friends.[14] And us, of course, the same as ever. We weren't going to Mars. That connection, those inalterable elements, are what creates a sense of continuity where the past and the future come together. This is how our memory is formed, like an unbroken line through the story we each create

---

14 The game here is not to remember my list, but to draw up one of your own. What are the first twenty things that would appear on your list of essential things that would never change, no matter where you might move?

about ourselves, like the ship of Theseus, which changes all the time while never ceasing to be the same.

When I emigrated as a child, the journey was very different. We rushed into exile urgently, with no time for contemplation. Amid the scramble, my mother explained innumerable things to us. I don't remember a single one, but I do know it was a trip filled with fear and doubts about whether my dad, who had left a few months earlier, would be there waiting for us. He was. Our flight was delayed by almost a day and there were boring layovers. Apparently, as we were later told, my brother and I staged a boycott and refused to continue. Why were we leaving what had been our home for four years? I forgot everything about that life that began and ended in the neighborhoods of Buenos Aires. There are photos, stories, names, even dogs. They are all someone else's memories. That trip fragmented my memory in such a way that the events on either side of the ocean were disconnected from each other.

My first memory is of the day we arrived at the house in Barcelona. My dad had put together a game with toy motorcycles on a track; they shot out like a flash and were magically suspended upside down in a double full flip. That is the starting point of my autobiographical memory. In that place and in that time, the stories that constitute my life begin to be knotted and linked together. Before that there is a great void, a space filled with other people's stories, collective memories. I was someone else. Almost all of us have a story that simple, although some of our first memories are much more glamorous or scatological.[15]

I think of my experiments on memory, decisions, learning, and emotions as ways to investigate the human condition. But actually

---

15 In 2011, Louis C. K. described his first memory to the Beacon Theater audience as follows: "How far do my memories go? [. . .] I was four years old. And I was standing in front of my parents' house and I was shitting in my pants. I was just shitting a massive, terribly painful shit. And I was halfway through the shit. That's my first memory, being halfway . . . The first half of the shit, I don't remember it. That's still in the ether of infancy. But the center of that shit was so wide that I actually came online as a result of the anal pain that I was experiencing. It actually awakened me into the stream of consciousness I am now living. That's how my life started. That's who I am."

I think I carry them out as a way to investigate matters that affect my own life. I have turned the questions we all ask ourselves into a profession. In retrospect, this has been the common thread in my research. The questions I ask are about those things I find most difficult to solve, that most hurt me or disturb me, the questions I want to study so I can find a better version of myself. What I've learned about our first memories has helped me to understand and build the story of my own life, to define who I am. I think it's an impressive tool for anyone wanting to shape their own story, which is why I've written this book.

Infantile amnesia

There are many ways to identify the memory from which the ball of memory can be untangled. Generally, it dates back to the age of three or four years old. The exact date varies, but it is extremely rare not to have a single memory of your first six years, or to have any of your first few months of life (which is highly paradoxical because there is no moment that is more transformative and defining of our identity).

In our first months of life we discover the universe: objects, people, and even ourselves. The brain, which weighs about three hundred

and fifty grams when we are born, multiplies its connections to triple its initial weight in three years. Therein lies the paradox, because the vast majority of adults do not have a single memory of those years when the mental and cerebral revolution that established them as individuals took place.

Freud made a fundamental thesis with this paradox and gave it a name: infantile amnesia, which today, after many years of accumulated science, is explained by two related principles. The first is the transition between different memory systems, a bit like what happened when people over forty made the leap to the digital world and lost the photos taken with analog cameras, or the fossil remains of Myspace, or what will remain from Instagram when it becomes obsolete in the future.

The second is that autobiographical memory needs the scaffolding of words and only acquires true meaning when one can share it with other people. There is a moment in childhood when experiences begin to be communicated spontaneously. Some things are told to others, and some things are hidden. That is the beginning of an editing process through which each person begins to build their own persona. That is where one's self-identity, one's self-consciousness, is consolidated. And, from there, memory finds a palace in which to organize all memories.

The easiest way to find out about people's earliest memories is by asking, which is what Freud did with his patients and what researchers today do with much larger samples. Those memories are usually quite stereotypical; the most common include toys (as in my case), our own homes, sudden frights, dreams, trips, vacations, and sibling births.

The answers are snapshots of children's memories seen from the perspective of adults and, as such, are highly prone to edits and distortions. So in order to accurately capture when our first memories are formed, we have to go back to childhood. It is also essential to specify which memory we are referring to, because knowing how to ride a bicycle is very different from knowing what day our birthday falls on, or that we don't like grapefruit. We won't forget any of those three things, so they are a part of our memory, but they are different kinds of memories. Neurobiologist Larry Squire made the first important subdivision of this taxonomy into implicit and explicit memory. Implicit memory is the sum accumulation of unconscious knowledge that forges our behavior: for example, learning to maintain our balance, breathe, and breastfeed. We learn each of these things without being able to explain how. They are built of perceptual and motor memories that we accrue from the first day we are born, or that are even part of the legacy of our species.

Explicit memory, such as the date of our birthday, is instead conscious. It comprises those things that we know we have learned and that we can explain. The question about our first memory is, therefore, an explicit memory that we can tell other people and that forms part of our conscious story.

Explicit memory is in turn divided into two broad categories: *semantic* and *episodic*. Semantic memory is all our factual information: who each person in our family is, what continent Canada is on, or that rain comes from the clouds. Episodic memory records what happens at precise times and in precise places. For example, the memory of a trip, a gift, or a kiss. The first times that Jacobo Bergareche references in his novel are archetypal examples of episodic memories. Episodic and semantic memory are interwoven. Neuroscientist Endel Tulving suggests that the best way to differentiate them is to think about how they are perceived: semantic memories produce a "feeling of knowing" and episodic ones a "feeling of remembering."

Psychologist Andrew Meltzoff showed that six-month-old babies are already capable of forming episodic memories; what they haven't yet developed are the capabilities that allow these memories to be lasting. The first is the capacity to have an autobiographical narrative to which they can attach this episode. When children begin talking about themselves, they use generic expressions such as *baby*, the wrong pronouns, or impersonal linguistic constructions. Only over time is identity clearly expressed through the use of personal and possessive pronouns like *I* and *my*. Only then do children begin to form accurate and lasting memories. The second missing capacity is language: our first memory linked to a concept is formed approximately one year after having acquired the word that designates it. That is, words are the underlying substance that stabilizes memories. The memories of the first months are unstable and vanish relatively soon, they remain in infantile amnesia's mnemonic cemetery, a broken and aimless ship unequipped with the anchors of words or the notion of identity.

On computers, interestingly, the main folder also alludes to identity. It is usually called "My Computer" or takes the user's name. There are other folders within it, containing different elements of identity: photos, videos, texts, documents. Those who maintain that structure can usually find their files. Those who, like me, keep everything on an overflowing desktop often suffer from electronic amnesia and lose their digital memories, which become files that are "on the tip of their tongue": they are somewhere in the mnemonic ocean, but there's no way to recall them.

We've seen that our memory is constructed through a creative process in which individual memories are woven together: like painting a picture in a memory palace or writing *the* novel of your autobiographical story. That is why language offers a lasting foundation for episodic memories. Words connect them into a continuous plot that makes up the story we tell ourselves. That is where identity is forged, just as when I concluded, after that race, that I was no good at sports. The memory of my vomiting was real; the rest was only a story I told myself. We will now follow this path, seeing how our autobiographical narratives blend with fiction. This means our memories are less precise, but at the same time offer us a powerful tool.

Illusions of memory

When we start recalling our first memories, there's always someone who claims to have an exceptionally precocious one. Some people evoke memories from the womb or from a previous life. As far-fetched as this idea may seem, if we dispense with the *declarative* element of memory, it starts to make some sense. There is room within us for implicit memories that are encoded in our genes.

The genome is a vast archive of information learned through mutations, selections, and adaptations. This genetic baggage that regulates the cellular machinery also triggers thought, from our very first day. Incredibly, a baby is born with basic concepts in mathematics, and moral and social bonds. In recent decades, scientists have shown ample proof of this genetic memory that configures and consolidates thought. The idea of the brain as a blank page, which had been paradigmatic for hundreds of years, was shattered.[16]

Harvard psychologist Susan Carey has captured this fabulous paradigm shift in her book *The Origin of Concepts*, which begins by presenting the different ways of investigating the infant mind. The most effective uses the expressive capacity of our eyes. We look—now

16 Deleuze would have had a field day.

and as babies—at the things that surprise us and trigger our attentional system. So by following babies' gazes, the layers of what they do and do not know are progressively revealed. Liz Spelke and Véronique Izard, for example, showed a series of images to a group of newborns: three dogs, three red squares, three large circles, three small sticks . . . After showing them that series, the children were offered two options: an image containing three objects and one with four. Babies will look much more at the one with four. This is not explained simply by a preference for images with a greater number of objects, because these experiments also included the reverse case: long sequences of four objects. Then, when the option between three and four is presented, the babies look more at the one containing three. A concept as abstract as that of cardinality is accessible to the brain of a newborn.

The machine that builds reality

Just as watching a baby's eyes is the best way to understand how he or she thinks, the worst way is to try to remember how we thought as children. We have already seen the reason for this: infantile amnesia. Our adult story is distorted by inevitable forgetting and editing. In *Conversations with Jean Piaget*, the famous child psychologist explains it with this story: "I was still in a baby carriage, taken out by a nurse, and she took me down the Champs-Elysées, near the Rond-Point. I was the target of an attempted kidnapping. Someone tried to grab me out of the buggy. The straps held me in, and the nurse scuffled with the man, who scratched her forehead; something worse might have happened if a policeman hadn't come by just then. I can see him now as if it were yesterday—that was when they wore the little cape that comes down to here (*he motions with his hand*) and carried a little white stick, and all that, and the man fled. That's the story." It seems like something out of a fabulous autobiographical memoir, except that Piaget himself points out a problem. "Then—I must have been about fifteen—my parents received a letter from the nurse [. . .] she had invented the kidnapping story herself [. . .] I must have overheard the story, and, starting from that, I reconstituted the image—such a beautiful image that even today it seems a memory of something I experienced."

As we saw earlier, our brains mix together fiction and reality, unbeknownst to us. And this happens in every realm of thought, not just in memory. I discovered this when I was living in New York, thanks to an experiment by Anne Treisman. A slide flashes on a screen, showing, on one side, a white triangle and, on the other, a yellow blot. When asked to describe what they've seen, participants instead simplify the image, declaring they saw a yellow triangle, something that was never on the screen. Perception, like memory, is more like a story than a reproduction; more like painting than taking a photograph.

When data is scarce, the brain constructs the simplest and most compatible explanation by composing a story with that information. This mechanism of unconscious inference is at the heart of visual illusions. For example, on the chessboard in the following illustration, squares A and B are identical in color. In fact, even though it seems impossible, the squares on the right are the same hue as those on the left. Here I show the squares out of context to reveal that they are indeed the same. The skeptical reader can verify this by covering the image on the right with a piece of paper, so that only those two squares are visible.

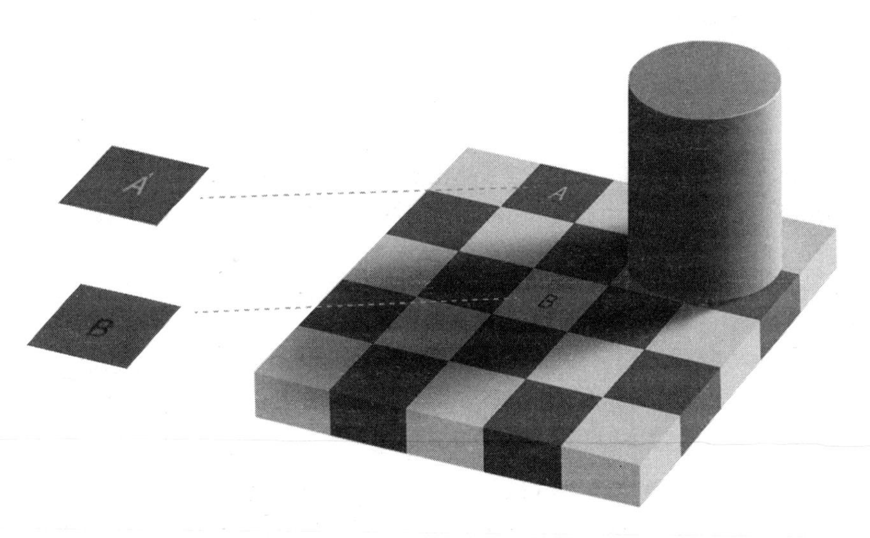

The unconscious brain carries out an impeccable logical exercise. Since square B is in shadow, it deduces that its true luminosity is greater than that recorded by the eye. And that's how we perceive it. We do not see the light that impacts our retina but the result of complex and sophisticated calculations made by the brain. The most curious thing about these illusions—as also happens with memory—is that even when we discover the trick, they are not dispelled. Seeing the whole picture again, the illusion persists: the brain, stubbornly, continues to construct its own reality.

The invention of memory

Treisman's experiment illustrates a fundamental element of our cognition: when presented with a triangle on one side and the color yellow on the other, we create a yellow triangle. The same goes for false memories: they are reconstructions that narratively connect *real* and unrelated memories in a plausible plot. We are amphibians with amphibious memories. With scarce information, the brain always constructs a basic and plausible story.

Our collective memory is filled with *yellow triangles*. One of them involves two legendary Argentine athletes: Diego Armando Maradona and Roberto Perfumo, that defensive center known for both his elegance and his roughness. In the book *Hablemos de fútbol* [*Let's Talk about Football*], Perfumo tells the same anecdote that the number 10 recalled in a famous television interview about the first time they faced each other.

In a clash, Perfumo sent him flying through the air, and then rebuked him with a smile: "You're totally fine, isn't that right, kid?" From the ground Maradona replied: "Of course, sir. And you? Is your foot OK?" The story of that kick was forged in the memory of both protagonists, discussed by them in different television programs and used as an example of tradition, of sporting respect, and of the tension that always exists between a consolidated legend and the up-and-coming genius. But the records show that Perfumo and Maradona never faced each other in a soccer match.

This fake anecdote is, of course, harmless. As too are false claims about yellow triangles on Treisman's screen. At the legal level, however, the same principle can and often does have catastrophic consequences. Witnesses can emphatically insist that they've lived through a certain situation even when it never happened. In her book *Eyewitness Testimony* (1979), American psychologist Elizabeth F. Loftus warned of the valid doubts raised by the statement "I saw it with my own eyes." It is a formula that is equivalent to saying: "Nobody told me what happened: my eyes photographed it and that is why I can offer you this portrait, which is the most reliable one possible." We now know, however, that such statements can be false, as much so as Piaget's attempted kidnapping or the impertinent reply Maradona gave "El Mariscal" Perfumo. These aren't exceptions or oddities: we all invent memories.

To see how we can use this idiosyncratic element of memory to our advantage, first we have to understand more precisely how memories are built and reconstructed, and where we have the opportunity to intervene to mold them as we wish.

Simmering on a slow boil

Over the course of a day we remember all kinds of details: where we left the car, what time we're meeting a friend, the cafe where we ate breakfast, and even the face of the person at the next table. Most of these memories fade quickly. All our school years can be described in a few hours of storytelling. The endless daily details that happened in those years vanish; that's how memory works. We also know, from experience, that the selection process is capricious: we remember the most decisive moments but also irrelevant episodes.

In order to understand the arbitrariness of memory, we will delve into the intimacy of brain function revealed through a series of experiments conducted more than a century ago. In one of those classic experiments, a sound is played (to a rat) and then an aversive stimulus is applied. The repetition of this sequence forms a memory that triggers a defense mechanism every time the rodent hears the sound. That memory lasts for many days; however, if shortly after that lesson is learned the rat is injected with a protein-blocking drug such as anisomycin, the memory fades away: the animal hears the sound as if for the first time.

By connecting one neuron to another, synapses form the cellular basis of memory. Anisomycin prevents new synapses from forming, but does not destroy those that have already formed. So what is strange then is that when the drug is administered after learning—when supposedly the new synapses have already developed and the memory has already been formed—the memory dies out. What's going on?

The answer to this question revolutionized how a memory is conceived. The synapses that make it long-lasting are not formed at the very moment of learning, but during the hours, or even days, after learning. This is what is called the *consolidation period*. Faced with the old idea of memory as a photograph, a new way of understanding how we remember suddenly emerged. Memories are simmered over a slow flame; only through consolidation do they become stable and able to resist the effect of the drug, precisely because bridges have already been formed. Remember: anisomycin does not destroy synapses; it prevents them from forming.

In recent decades we have discovered that, during consolidation, new synapses articulate a well-defined brain network. When memories begin to form, a structure known as the hippocampus plays a part. The other celebrated role of this brain region, as discovered by Norwegian Nobel laureates Edvard and May-Britt Moser, is forming spatial maps to orient ourselves and navigate through space. This makes sense because, as we know, space is memory's natural grid.

We can think of the hippocampus as a system of indexes that links the different attributes of a single memory: a sound, an image, a place, an emotion. Each of them is encoded in neural circuits that are in different regions of the cerebral cortex. If we form a memory that brings them together, their circuits are connected by synapses. This connection is vulnerable and would break if not for the work of the hippocampus, which consolidates these cortical connections into a process of mnemonic reverberation. It is like when we repeatedly recite what we want to remember, but in the privacy of the brain, without our knowing. To accomplish this task, the brain requires silence or, better yet, sleep. When the cortex is freed from having to process external stimuli, it can devote itself to consolidating memories, led by the hippocampus. That is why sleep is the vital fuel of memories. Without it there is no reverberation nor consolidation, leading any acquired memories to fade and sink into the great swamp of oblivion. Reverberation strengthens the synapses connecting the circuits through which different attributes of a memory are encoded. When the memory is consolidated, connections between cortical regions become durable and autonomous, freed of the role of the hippocampus. The circuit of neurons that encodes all the attributes of a memory is called an engram. When the neurons of an engram are activated, we recall that memory.

Earlier we asked ourselves what happened to the places, people, ideas, movies, and books we fleetingly remember. Now we know: they were never consolidated. While we amass memories at a dizzying pace, only a few are consolidated, in a slow and laborious process that requires hours of calm and silence in the hippocampus and cortex.

Ode to oblivion

We are still missing a key piece in the memory puzzle, one that was discovered by James Misanin at Rutgers University. A memory can be erased if protein synthesis, and therefore synapse formation, is blocked at the precise moment it is being recalled. In other words, when a memory "comes to the surface" it becomes as fragile as it was before consolidation. This is the moment when it is being reconsolidated.

Neural reconsolidation is easier to see through an analogy with the digital world. Let's use as an example a text file on a computer's hard drive. This file is a lasting memory but it becomes vulnerable the moment we open it and make it editable. At that point, anyone can add text, delete or change what was already written. When we "save" the document again, either voluntarily or by mistake, this new version replaces the previous one. The same thing happens in the brain: a memory becomes modifiable when it is recalled. New synapses can be added, existing ones can be deleted, their attributes changed. Then the brain reconsolidates that memory. It's the equivalent of pressing the "save" button. And so, memories mutate with each revision. Before we look at how this web of edits and reconsolidations gives rise to false memories, let us continue with our journey along the cellular level to analyze the most drastic case of editing. The place where almost all memories end: oblivion.

Why is it so hard for us to rid ourselves of some memories, even when we try with all our strength? Could it be that they become progressively more persistent the more times we revisit them? As intuitive as that idea is, it is also erroneous. Hal Pashler, of the University of San Diego, demonstrated this by studying how long it takes us to forget a memory as a function of how often we recall it. It turns out that evoking something very insistently, day after day, creates an ephemeral memory. Who among us can remember the things we reread countless times in preparation for a school exam? If we review a memory in daily sessions, it will fade on the same scale of days. If we revisit it once a week, it fades on a scale of weeks. If we further space out the time between sessions, the memory lasts progressively longer. In other words: if what we are trying to do is create lasting memories of what we've studied at school, less is more.

So the question remains. Why are some memories so hard to forget? The key lies in the nature of oblivion: memories are lost when they are disengaged from the recall apparatus. Nobel laureate Susumu Tonegawa proved this in an experiment that showed it is possible to evoke an extinct memory by stimulating the set of neurons that comprise its engram. To put it another way, a memory does not completely disappear when it is extinguished during reconsolidation. What is lost is our ability to evoke it, like the words that remain on the tips of our tongues. If lost memories are those that are disconnected from the evocation system, it is reasonable to assume that, on the contrary, the unforgettable ones are more connected to it. To visualize this idea, let's return to our digital analogy.

On the internet there are also memories that are almost impossible to erase, which has given rise to an ethical, technological, and legal movement that defends the *right to be forgotten*. These stubborn memories are those that are indexed with many words and high priority. The image of Google is often mistaken for the memory container, but it is not. Google works exactly like the evocation system of all digital memory. The famous algorithms of Spotify, YouTube, Facebook, and Google do not regulate how indelibly a file is recorded on the hard drive, they merely change how easily we access it. The same goes for memories: they are persistent and unforgettable when they are connected to the recall system by many branches, when they appear in the playlists of our brain's algorithm.

The imprint of an emotion

We know now that the most difficult memories to forget are those with many pathways to recollection. When does this happens? The intuitive answer here is the correct one: the key is emotion.

The most emotion-packed moments usually generate memories that we, like an elephant, never forget. Those of us who were already adults on September 11, 2001 when two planes crashed into the Twin Towers, will remember it vividly. We recall not only the images of the event, but also every detail of our lives on that day: where we were, with whom, what we were doing at the time. Each of these elements, encoded in different circuits of memory, are united in an engram that links that entire experience with other emotional circuits in different regions of the brain. A color, a smell, a place, an emotion, or an image is enough to activate the evocation system. Stressful memories are difficult to eradicate because they are linked by thousands of connections and associations.

The Argentine neuroscientist Pedro Bekinschtein revealed that the key to this phenomenon is cortisol, a hormone produced in response to stress. When cortisol is pharmacologically inhibited, memories become more specific. An association between two stimuli is learned, but this association is disconnected from the context, which prevents its generalization. By way of an example, let's say someone has a traumatic experience with a dog. Normally, this experience will become generalized, increasing our aversion to all dogs, and perhaps other animals as well. When cortisol is blocked, however, the memory will be associated with the dog in question or even just some of its actions. The memory will be formed, but it will remain in a susceptible state. Memory without cortisol (without stress) has few triggers that connect it to the evocation system, and that makes it fragile.

Let's look at the evolution of these ideas in the medical field. Post-traumatic stress disorder (PTSD) is chronic suffering caused by the persistent memory of a painful event. It can be brought on by a wide range of events: some extraordinarily violent ones—such as a kidnapping, an accident, or a rape—but also by other much more common ones, such as infidelity or purse snatching. These incidents leave lasting traces of suffering and latent depression. Post-traumatic stress is seen in approximately 70 percent of the population. Everyone has their own issues.

The main tool psychotherapy uses to deal with post-traumatic stress employs the process of reconsolidation to its advantage: by evoking the traumatic memory in a safe, relaxed situation, in order to begin to forge other associations and move the engram towards less traumatic regions. During the evocation is when words are most effective at reconstructing a memory. This technique is quite effective with memories of a moderate emotional charge. In the case of the more traumatic memories, however, the process of trimming and mending is much more complex and a combination of words and various drugs is being tested for its efficacy. This may seem like a modern idea, but it's not. As far back as Homer we have accounts of potions that facilitate very painful conversations. When Telemachus arrived at the wedding of Menelaus's children, he did not know whether his father, Ulysses, was alive or dead. The conversation about the Trojan War in the middle of the occasion was both painful and necessary. To make it possible Helen mixed *nepenthes* into the wine. *Nepenthes* means "painless": "The man who has drunk it, mixed with wine in the vessel, will not shed tears on his cheeks for all the day."

History repeats itself more than two thousand years later. In 1912, the German chemist Anton Köllisch synthesized MDMA in the laboratories of the pharmaceutical company Merck, where he worked at the time. The substance went under the radar until 1970, when the chemist Alexander Shulgin rediscovered it and studied its possible complementary use in therapy for post-traumatic stress. By then, MDMA had already become famous for its recreational use. Soon after it was banned in almost every country in the world and research on its therapeutic use was stopped in its tracks. It took forty years for Michael Mithoefer to be able to resume the project in California and discover that, in severe situations of post-traumatic stress, MDMA-assisted psychotherapy is effective and safe. There they were, Helen and her *nepenthes*, making possible those healing conversations that, in the presence of tremendous pain, seem impossible.

Even more frequently used in the treatment of post-traumatic stress is propranolol, an adrenaline inhibitor that, like the cortisol in Bekinschtein's experiments, mediates the stress response. The results of a meta-analysis of clinical trials show that this drug decreases stress only if it is administered during the reactivation of traumatic memory, acting at the time of reconsolidation. It is ineffective at any other time: the threads of a memory can only be cut when it is evoked.

Clinical post-traumatic stress is the exaggerated version of something much more common. We all, to a greater or lesser extent, have toxic memories that plague us, and we all would like to, at least partly, calm that suffering. A general technique for regulating emotions is to observe them from a distance, lowering our vigilance and reactivity, as if they were being experienced by another person. That mental state causes cortisol inhibition. The brain doesn't differentiate between cortisol inhibition provoked by a drug and cortisol inhibition provoked by a conversation: the healing effect on emotions and memory is the same in both cases.

True memory is fake

Connect the dots was a game found in old children's magazines. The idea was to draw a line between the number points, from 1 to 2, then from 2 to 3, and so on, until the cloud of isolated dots reveals an image. All the information to form the image is in dots, but in a cryptic way that is only revealed when they are connected by pencil lines. Treisman's illusory conjunctions also work by connecting dots. Our brains link the yellow blot on one side of the image with the triangle on the other, and recall a yellow triangle.

Connections of this kind appear in various fields of perception. This is the case with an auditory illusion known as *the phantom fundamental*, used centuries ago for musical composition. When you play a note on an instrument, what actually sounds is a stack of notes known as harmonics. For example, when playing an A note on a guitar at a frequency of 440 Hz (that is, the string vibrates 440 times per second), other frequencies also sound, multiples of the fundamental. Double (2 x 440 Hz = 880 Hz), triple (3 x 440 Hz = 1,320 Hz) and larger multiples. Other small fractions of the fundamental also appear, for example 3/2 x 440 Hz = 660 Hz. This is not a side effect: it is that richness that defines the sound of an instrument. Some, such as requinto guitars, even have strings that are never played, whose only purpose is to change the sonority of the harmonics.

The rule of harmonics gives rise to a mind trap. If we play frequencies of 220, 330, 440, 550 and 660 Hz at the same time, the brain processes them as coming from a fundamental sound of 110 Hz, of which they are all harmonics. That is to say, even when the string is not vibrating at that fundamental frequency of 110 Hz, we hear it, even more clearly than those that actually sound. This is the effect of *the phantom fundamental*, by which we hear a note that is never played. Once again, the brain unconsciously processes sensory information by connecting distant points with a sophisticated logic that builds perception.

This effect that regulates our auditory world has its counterpart in the realm of words and memory. What word comes to mind when reading *horse* and *stripes*? Almost everyone thinks of a *zebra*, a middle point (dot) that connects those two distant concepts. Henry Roediger and Kathleen McDermott demonstrated in their very famous study that this connection mechanism is a fundamental tool of memory. The iconic experiment works as follows: participants hear a list of words, tasked only with remembering them. The list is chosen very carefully in order to form a semantic cloud around a concept. For example, *bed, tired, waking, blanket, snoring, pillow, relaxation*, and *yawn*, which form a cloud around *sleep*. Some time after they are presented with this list, the participants remember hearing *sleep* with much greater probability than some words that were actually on the list. *Sleep* is like the phantom fundamental: a word that doesn't appear on the list, but that is the source of all the words on the list. Even though it isn't there, we remember it. The brain connects the dots and, in doing so, "remembers" the spaces between them. In memory, as in perception, what remains are yellow triangles.

The false memories in Roediger's experiment are not arbitrary. The distortions are limited to neighboring words: we think we remember *sleep* because it is close to *bed* and *snoring*, *chair* because it is close to *table* and *armchair*. Let's imagine a game that repeats this process over and over, sort of like the game of telephone. A new word is produced from a list of words. After several rounds, we will have extremely sophisticated lists, which have traveled far from their starting point. Here we have an analogy between the world of neurons and the world of words. Each round of the game is a moment of evocation and reconsolidation when the memory engram is edited, changed slightly and recorded again in this different version. The repetition of this process generates a series of very elaborate memories filled with more and more imaginary details, like those of Piaget, or Maradona and Perfumo, or those identified by Elizabeth Loftus, which produced fabulous testimonies that put a monkey wrench into the judicial system. Memories are formed and deformed along the same brain circuits. This is what makes true memories almost indistinguishable from fake ones.

False memories are often considered an error. Like any file system, distortions occur in human memory both by omission (what is not remembered) and by construction (what is remembered, but never happened). The results I've presented here suggest a very different idea: transforming memories is a kind of freedom that gives greater malleability to the creative use of our memory. False memories—which perhaps we should more properly call *illusions of memory*—are the result of a sophisticated and creative system: linking each memory to a story that gives continuity to our identity.

The creativity of false memories

The *modus operandi* of false memories suggests their connection to creativity. But being able to clarify this relationship requires a good tool for measuring creative thinking. In 1960, Martha and Sarnoff Mednick offered us one: the Remote Associates Test (RAT), which evaluates a central facet of creativity: lateral thinking. The test consists of finding one word associated with three others, for example, *apple, family,* and *house.* There is no single solution to the problem and therefore the test cannot measure creative ability with absolute precision, but even so the RAT is a good approximation and has become standard. The solution to the example presented is *tree,* which is linked to each of the words as follows: "apples grow on trees," "the tree house, where I played as a child" and "the family tree that shows the generations." The RAT is a good game to play at home, to draw attention to the intricate way we store words.

The RAT test's mechanism works in a very similar way to that of false memories. Both involve the brain process that identifies the *phantom fundamental* in the world of words. So it should come as no surprise that inducing false memories—*à la* Roediger—turns out to be an effective way to enhance creativity. The loss of focus inherent in the search for concepts that is typical of mnemonic illusions has a *positive side*: it helps us to find creative solutions. And it is also helpful for recording, in the log where we write and read our saga, a better version of ourselves.

When we move to a new place we often face questions such as "Why did you leave?" and "What brought you here?" Any answer to these questions is somewhat deceptive, however honest our intentions may be. What stands in the way of complete honesty is our inevitable perspective from the present moment: we cannot return to the shoes of the person we were when we decided to leave—the only person who could answer those questions truthfully. As Carl Sagan wrote, "When we are asked to swear in American courts of law that we will tell 'the truth, the whole truth, and nothing but the truth,' we are being asked the impossible."

Perspective distortion is one of the signs of mnemonic illusions, sometimes even of the most absurd. For example, in one of his studies, Nicholas Spanos recounts the case of a person who claims to have lived in the past, and from the perspective of that previous life describes events that happened in 500 BCE. How could he be aware of Christ five centuries before he was born? The incoherence of this story exaggerates something that is characteristic of every memory: it consistently merges with the present, the only perspective from which we can genuinely tell our own story.

If the present interferes by continually distorting the story of the past, how is our identity constructed? Anne Wilson and Michael Ross solved this apparently unsolvable riddle: how can we change all the time while never ceasing to be ourselves?

Before we get to the experiments, let's move this idea to another realm where the tension is even more striking. Over the course of a few years, virtually all the matter in our body is replaced. Our skin, blood, bones, liver, and most other organs will regenerate at their own pace, recomposing with different atoms from the air we breathe and from what we eat and drink. In a few years we will be comprised of rejuvenated matter. However, we will still be ourselves. This paradox, relatively unknown, is very disturbing. Each person is a ship of Theseus. And that disquiet carries over into the domain of memory, thought, and consciousness: everything changes and nothing changes.

The solution is that we reconstruct the past from a moving perspective. Just as when you travel by car you get used to coasting and assume you're moving forward. Let's take a look at the experiments, which are revealing. In the first, professor emeritus of psychology Michael Ross tested a group of students, then taught them a course and, upon completion, re-examined them. Attending the course led the students to believe that their grades on the second exam would be better than those in the first. What Ross did not mention is that the course taught absolutely nothing that could improve students' performance on the assessment. How do you regulate the tension between an expectation of learning and the reality of a course in which nothing relevant is taught? The brain solves the problem without questioning the model of progress by creating a false memory that gives coherence to the story. When questioned about their initial assessments, the students recalled getting worse grades than they actually had. Those who'd gotten an eight before starting the course remembered getting a seven; those who'd gotten a five were convinced that they'd received a four: the false memory maintained the illusion of progress. Ross repeated the experiment with another group to which he also gave two exams, but without teaching a course in between. These students were able to accurately indicate the grades they'd been given in the first evaluation. So we see that false memories are not merely distortions. They are, in fact, akin to visual illusions: an unconscious and creative mechanism to make sense of scarce and contradictory information.

After conducting a long series of experiments with a wide range of participants, from students to great athletes, Ross came to the conclusion that devaluing the past to exalt the present is a very prevalent bias. To illustrate it, he quoted a passage from the autobiography of the Hungarian writer Arthur Koestler: "The gauche adolescent, the foolish young man that one has been, appears so grotesque in retrospect and so detached from one's own identity that one automatically treats him with amused derision. It is a callous betrayal, yet one cannot help being a traitor to one's past." Traitors to our past: that's the key. Just as jealousy and envy lead us to create unfavorable versions of others to feel more valuable in comparison, the desire for progress distorts the past until it becomes an unfavorable and grotesque version of what we once were. That, according to Ross, is what happens to almost all of us.

The clearest counterexample of this perspective is the bias that appears as we near old age. In this case, the reverse phenomenon occurs: in contrast to our decline we fabricate memories of a more exuberant youth than we actually lived and our present begins to live off our past glories. As the saying goes: everything was better in the good old days.

Each of us occasionally falls prey to these biases. Perhaps it is Nestor Burma, Léo Malet's celebrated detective, who best reflects our need to reaffirm the present while being compassionate and tender with the past. Accused by a policeman of having been an anarchist, Burma responds with the following quote from a French prime minister: "Any man who wasn't an anarchist at sixteen is an imbecile"—then adding—"just as much as if he remains one at forty."[17]

I identify with both perspectives in my own story. In the realm of skills, where Ross focuses his experiments, I agree with the rule. I learned to play the guitar and everything I know about music in recent years. As I mentioned in the prologue, I recently started cycling and now I bike thousands of kilometers through the mountains. I can recognize that, in both cases, I constructed a narrative in which my musical and sporting pasts were worse than they actually were, allowing me to applaud my newly acquired skills to myself and others. On the other hand, as far as my travels and engagements around the world are concerned, I believe that I'm elevated by a glorious reputation that my past might not live up to and that, I realize as I write this, is hostile to my present.

Identifying the arenas in which one perspective or the other prevails and the implications this has for our lives is a good exercise. The way we narrate our own memories affects how we experience them and shapes our emotional profile.

---

17 I read this comic about twenty years ago. When I went to find the exact quote to transcribe here, I found that . . . my memory had edited it. Burma does say the first sentence, but not the second, which is actually the policeman's sarcastic reply. I prefer my version, with my nostalgic view of a bittersweet Burma, and so I'll leave it as it is.

Exercise: Ideas for living better

1. **Ask (yourself, too!) specific questions**
   Don't ask your child, "How was your day?" or a friend, "How's your life going?" Overly broad questions lead to blocks, to the terror of the blank page. You're much more likely to discover something about your kid if you ask, "What was your last class of the day?" Your child will easily find a memory station and something interesting may arise.
2. **Don't expend your energy on trying to optimize impossible decisions**
   Pondering the options we have and explaining them to other people is fine, but when all the alternatives are equivalent it's reasonable to accept random chance and just pick one. In such cases, hesitation and floundering lead only to unnecessary suffering and a waste of your time.
3. **Aspire to deep learning**
   Think about something you are good at, something you know in full detail, something you can connect with other areas, some knowledge you feel confident and secure about. These parts of your memory are the most useful. They are sources of creativity and the freedom to think and reason better. The antithesis of this is inert learning, which floats on the surface and hardly connects to the rest of our knowledge and experiences.
4. **Train your memory**
   People who remember many things don't have larger compartments where they store more information: rather,

they are experts in creating stories and images to coherently connect the different pieces of their knowledge, so they are able to recover them in the midst of all the noise. You can do it too. Knowing how to read and write in your memory bank like this is one of the keys to thinking freely.

5. **Remember that the memory is a frame, not a photo**
   Memories change every time they are evoked, in a never-ending process of editing and correcting our own history with the objective of creating a coherent story. Sometimes, we even create false memories because they fit better with the narrative we tell ourselves. This ability, usually perceived as a defect, fuels creativity.

6. **Memory shapes our identity**
   We all create fables that alter our past to give meaning to our present. Sometimes we err on the side of naivete (remembering a splendid past that maybe wasn't so splendid) and other times on the side of severity (look how much wiser we are compared to our past selves!). Think about where you are too innocent and where you are too hard. Knowing that, and knowing that words offer us the opportunity to choose sides, is key to emotional self-care.

CHAPTER 4

The Atoms of the Mind

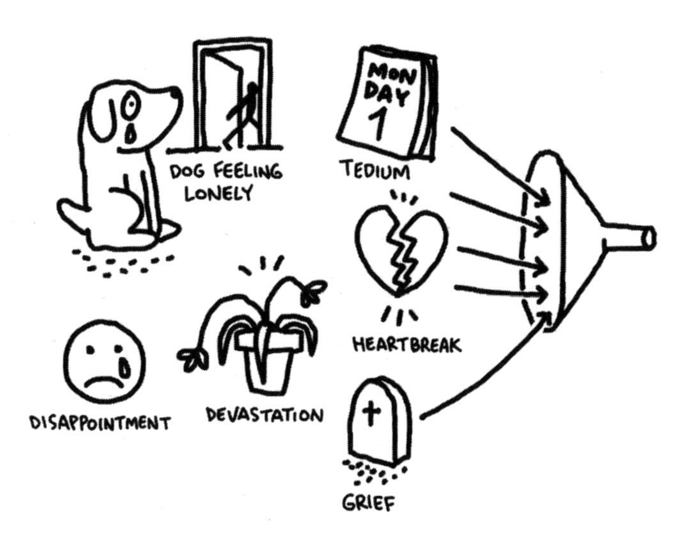

How to clarify what we think and feel

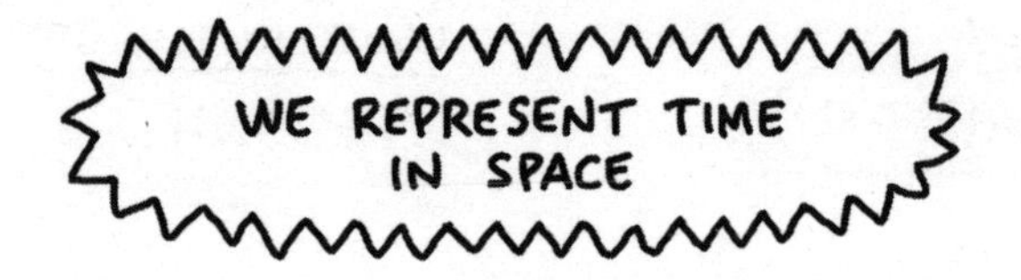

AND THE ABSTRACT IN THE BODY

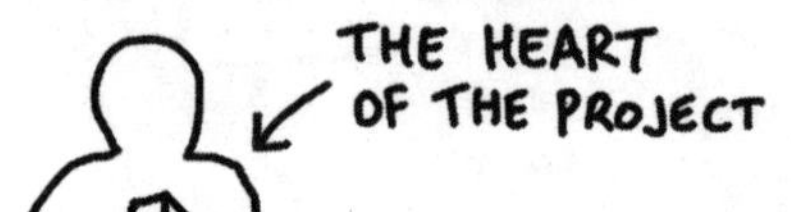

THE HEAD OF THE GOVERNMENT

THE LONG ARM OF THE LAW

THE FOOT OF THE PAGE

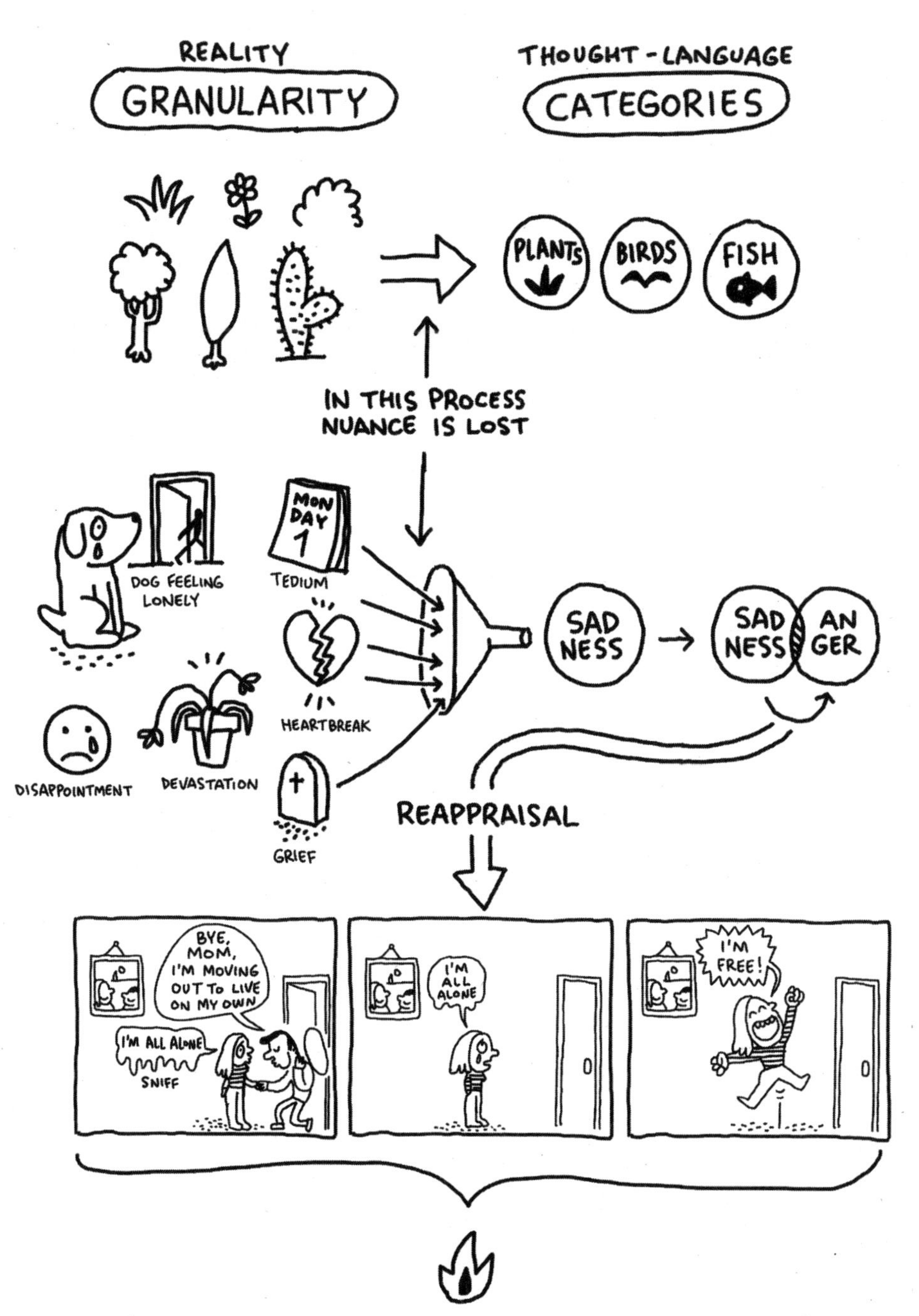

WHEN WE REAPPRAISE AN EMOTION, IT DOESN'T SWITCH OFF: WE JUST INTERPRET IT DIFFERENTLY AND EXPERIENCE IT IN A DIFFERENT WAY

Words are the building blocks of conversation. They inevitably transform what we feel: a single word can seduce or destroy, define the course of a negotiation, heal or cause pain. The perfect word is our greatest power or, as Dumbledore would say, "our most inextinguishable source of magic."

The link between language and cognition has been a favorite topic and latent debate in experimental psychology. Can ideas exist without language? How are those ideas transformed when described in words? We will see that language, without warning, gives shape to our thoughts. We will discuss abstract realms, such as time and space; others, such as smells, that seem impossible to describe in words; and also emotions.

Language is a double-edged sword. For example, compare what we experience when seeing a painting, giving a kiss, eating something delicious, or smelling the perfume of someone we love with the description of those experiences. Our descriptions allow us to project our experiences into the minds of others, but at the cost of losing nuance and shades of gray, both for others and for ourselves. That is the B-side of words: the world as seen through them is pixelated, like in Minecraft.

We represent our emotions with a broad brush because we have very few words to define our vast spectrum of feelings. The sadness of a child who's dropped their ice cream, the sadness of a sports fan whose team has just lost, and the sadness of a loved one's death are very different emotions. But we use the same word to describe them all: sadness. Since language is reflective, that confuses us and ends up becoming pernicious. We can also see the same idea from the opposite perspective: while language has its limitations, it offers us freedom. Emotions are often ambiguous and we have a lot of margin for reinterpreting them, redefining them. Some of us call a racing heart and butterflies in our stomach *fear*, while others call it *enthusiasm*. And this is not merely a semantic question: it drastically changes how we experience that situation and what we do as a result.

A while ago I read a flurry of tweets that were written by a very dear friend of mine from his home in New York. None of them was overtly worrisome. But there was something strange and disconcerting about that barrage of text that made me assume my friend wasn't OK. How do these sorts of intuitions arise? How do we read between the lines, beyond the explicit meaning of each word, to infer emotions in the mind of others? The brain is extremely effective in making this type of inference, so much so that sometimes it goes too far, drawing premature conclusions that result in stigmas and prejudices. The algorithm the brain implements to produce these intuitions is nourished by a simple ingredient that is at the heart of artificial intelligence: induction. It also allows us to discover thousands of words without anyone teaching them to us, and to get along in a world where things look alike, but are never exactly the same.

That day, reading my friend's tweets, intuition worked: my brain's algorithm detected a camouflaged alarm signal; I understood that he needed help, and I was able to offer it to him. However, many times we do not detect a cry for help and we miss the opportunity to lend a hand at the right time.

I found this idea summarized in the last story in *Extraordinary Tales* by Borges and Bioy Casares, "The World is Wide and Strange," which has stayed with me, almost obsessively. The story, just one sentence long, reads as follows: "They say that Dante, in chapter 40 of *La Vita Nuova*, says that when traveling through the streets of Florence he was surprised to find pilgrims who knew nothing of his beloved Beatrice." That infinite ardor we feel inside can remain completely imperceptible to others, even those closest to us. My life's whole adventure in science is, in a way, a means of bridging this gap. I suspect that, somehow or other, we all share this impulse. It is the impulse behind our laughter, caresses, hugs, our love. It is also the impulse behind our words, which have the fabulous ability to make the world less wide and strange.

Words and ideas

In Albert Einstein's theory of relativity, space and time are inexorably related and interchangeable because they are, in reality, different expressions of the same thing. This idea was a conceptual revolution in our understanding of the cosmos, but somehow it was already part of our collective common sense. When we say, "Christmas is coming up fast," where does it come from? From the south? From the east?

The verbs associated with space merge with those of time in a peculiar way. Sometimes the future shifts to where we are, as in the famous slogan of *Game of Thrones*, "winter is coming." Other times, we are the ones who move towards the future, as in the ineffable political slogans that suggest we "walk together to a better tomorrow."[18]

Whether we are moving toward it or it is approaching us, the future—both in language, and in our minds—is in front of us, and the past is bringing up the rearguard. We are invited to *leave* the past behind or to be hopeful about the future *ahead*. This belief, which seems unquestionable, is expressed in other ways as well. We'll extend our arm forward to refer to the future and point backward at the distant past.

Chronobiology researcher Juliana Leone, artist Mariano Sardon, and I conducted an experiment in which we asked each participant to draw three circles to represent, respectively, the past, the present, and the future.

18 History is important, we are told at school, in order "to know where we come from and where we are headed."

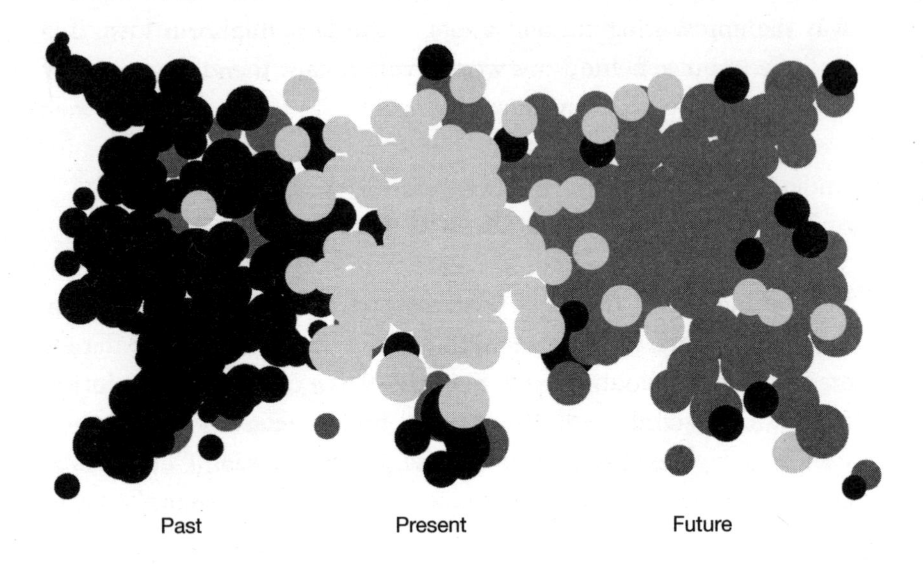

Each person placed the three circles wherever they liked on the page, revealing substantial differences. Some conceive of the present as tiny and their pages (and their minds) as filled with the past and the future, while for others the past and the future are tiny circles, orbiting, or sometimes contained within, the present. This variability persists within a common rule: the past is on the left and the future on the right, at least for people who write and read in that direction. Yet the rule that the past is behind us and the future ahead of us does not seem to be linked to any particular cultural expression, leading us to suspect that it's universal. However, it is not.

In the Andes region of South America, the Aymara people perceive the association between time and space differently. Carlos Núñez, a professor of cognitive sciences at the University of San Diego, explains that, when speaking about the future, the Aymara accompany their words with an arm extended backwards. The further the future to which they refer is, the more pronounced their gesture backwards becomes. And when they speak about the past, they extend their arm out in front of them. This way of thinking about space and time is based on their different use of words: in the Aymara language, *nayra* means "past" and also "in front" or "in view." And *quipa* means both "future" and "back." These words define another way of representing time in space, through metaphorical use that links the seen with the known. You see the known and the unknown is not visible. We use this metaphor all the time, for example when we say, "Do you see what I'm saying?" to ask if our interlocutor understands us, if we were *clear* in our explanation.[19]

The Aymara associate the past with the known and, therefore, with what is in view, in front. Since the future is unknown and out of sight, it is behind them. This logic is so impeccable that, when we hear it for the first time, we are tempted to start using it ourselves. After all, we know more about our past than about our future and this association with the visible and the invisible is easy for us to make.[20]

Not only the direction of time but also its geometry changes between cultures. For the Aztecs, the arrival of the Europeans meant the end of one cosmic era and the beginning of another: time for them was therefore circular. The bond between time and space is a cultural construct forged in language. This example illustrates a more general principle: *many realms of thought can be reappraised, even those that seem impossible to transform.*

The shape of sound

Time isn't only represented in space; it also has intuitive and automatic links to sound. In music, we go from low to high sounds by *raising* the pitch, changing the *height* of the sound, and this spatial association extends to writing: the bass notes are written lower than the high notes on the stave. This relationship is also expressed in the body; we usually lift the body when singing high notes and drop it slightly for the lower ones. This can be counterproductive because the low note can disappear. To remedy this, a typical exercise involves reverse mimicry: lifting the body or raising the arms to sing the

19 Then there is the famous quote from Saint-Exupéry's little prince: "The essential is invisible to the eyes." What else could it be invisible to? Clearly, to all that is known.

20 "The future lay before him, inevitable but invisible," says John Green in the novel *An Abundance of Katherines*. The word *before* frequently means "earlier" but here it is used in the sense of "ahead."

lowest notes and lowering them for the high notes, using the body to reverse the stave's order. It is a way of reappraising sounds, of changing the stereotyped associations between high frequency and high energy, revealing that this link, despite how intuitive it is, can also be shifted.

Language also establishes links between forms and sounds that can be as intense as they are difficult to explain. The most famous example comes from an experimental test question as simple as a marketing survey. It was conceived by my friend and colleague Edward Hubbard along with Vilayanur Ramachandran and its success has made scientific celebrities of these two shapes: one is called Kiki and the other Bouba. The question is: which one is which?

This experiment has been repeated over and over again, with people of all ages, and across different countries and cultures. The consensus is nearly unanimous: the illustration on the left is Bouba and the one on the right is Kiki. The association seems as obvious as it is difficult to explain. It is like when we say someone "looks like" an Amy, or a Carlos, which is never really the case. It turns out that with Kiki and Bouba there is a good way to explain how the unconscious mind of (nearly) all of us arrives at this conclusion. When pronouncing the vowels /o/ and /u/, our lips form a circle that corresponds to Bouba's roundness. But to pronounce the /k/, the back of the tongue rises and the palate closes in an angular configuration. That is why the pointed shape of the illustration "looks like" a Kiki.

The examples of time and space and the Kiki and Bouba test show how *language conditions our way of thinking, from the most abstract ideas to seemingly irrelevant decisions.* They both reveal the procedure that gives rise to these beliefs and which is, in turn, the starting point for reconstructing them.

In Aymara and in English, the metaphors that connect time and space are different. When and how did these timelines fork? How do concepts change when they are named in words? The key is understanding that a word gathers together a group of different things that have something in common. The word *dog*, for example, defines animals of varying heights and weights, young and old, and a wide range of colors. It can even refer to drawings or paintings. All of these instances are part of the *dog* category.

Language sounds are a good example of how a continuum fragments into categories. French vowel sounds can be hard for English speakers to differentiate, and vice versa. The phrase "*les jeunes gens jaunes*" sounds to us like the repetition of three identical words: "*lay shon shon shon*." To a native French speaker the vowel sounds in these three words are completely different, and utterly unambiguous; *jeunes* means young, *gens*, people, and *jaunes*, yellow. There are worse cases, "*cou*" and "*cul*" are often heard as the same—as if both were "*cu*"—but the sounds are actually very different, and confusing the two can get us into hot water: one means "neck" while the other means "ass." Why are we unable to distinguish vowel sounds that a native French person easily can? Patricia Kuhl proved that it is a lost skill; every baby, anywhere around the world, can differentiate the sounds of all languages, including "*cul*" and "*cou*" and "*les jeunes gens jaunes*." What happens is that this ability declines as those babies are exposed to their culture, and finally disappears at puberty.

To understand why we lose this ability, it is helpful to identify the continuum of sounds. We can do this by placing our mouth as if about to say /e/ and then say /a/, so we produce a sound that is right in the middle of these two vowels. With a little practice you can turn one vowel into the other in a continuous and progressive way. When we acquired the sounds of language, long before we produced our first words, our *job* was to parcel out this continuum of sounds into just five vowels. The *job* of someone who grows up listening to French is to divide it into a different set of parcels, and so each culture chooses how to divide the sound map into categories.

It turns out that to learn these categories you have to unlearn the differences between sounds within the same category. My /a/ is different from my friends' /a/, and from my siblings'; different even from the /a/ I pronounced as a child; different from the /a/ I pronounce first thing in the morning and then at the end of the day. So we must learn that all of these sounds are different instances of the same category. And as we do that, we lose resolution. The sonorous objects we could once distinguish begin to be confused within a category. And it turns out that several parcels of one language can fall within a single category in another language: like the sounds of "cou" and "cul," which are different, but fall within the category of the vowel of our /u/ and as such we cannot hear the difference between them. What is different in the sound is heard as the same in speech.

What we've just seen with phonemes also happens with all thought, even in those corners of perception that seem less likely to be labeled with words, such as the seemingly indecipherable world of smells.

Fruity and nutty

As we walk around, we are struck by an unexpected aroma and suddenly transported to our childhood. We are suddenly trapped in a torrent of unclassifiable evocations, as incredible as they are inexplicable, similar to Proust's madeleine, which was able to abruptly and unconsciously unleash a stream of memories that had been buried for years.

Smell is one of the best examples of a wordless mental experience. That is why, in his ode to rationality, Kant considered it the most expendable of the senses. The German philosopher said that smells can only be described by referencing something outside of the olfactory realm. The words designating smells, in most languages of the Western world, usually refer to the substances from which they emanate, such as the smell of vanilla or coffee. This designation has an obvious problem: most scents result from combinations of materials and cannot be described algebraically by their components. A wine-tasting note might read, for example, that "in the nose we find varietal typicity with notes of pepper, jam, and a wood finish." But the aromas that populate our lives create a more complex and indivisible orchestra: mixtures of the smell of factories, smoke, people, trees, rain, and earth. The world of smells, except in exceptional cases, is too intricate to separate out.[21]

That conclusion is a cliché. It is part of our intuition, part of our philosophical and scientific discourse. The problem is that it derives from a very unrepresentative sample of humanity: Western, educated, industrialized, rich, democratic individuals; what Joseph Henrich, Steven Heine, and Ara Norenzayan called, in a play on words, "the weirdest people in the world" because of the acronym formed by that list of adjectives. But the world is wide, and contains many more people than just the *weird* set; many anthropological studies have identified cultures whose languages are rich in words to describe smells.

In the past decade, Asifa Majid, a linguist and psychologist at the University of York, has worked to demolish the myth of this divorce between smell and language, which was built on the partial observation of Western societies. In one of her pioneering studies, she looked at the language of the Maniq, a tiny population of nomadic hunter-gatherers in southern Thailand. The Maniq's language (and this is not anomalous) has about fifteen words that refer to smells.

---

21 Getting into a taxi in Buenos Aires, you may find yourself shocked by its *sea breeze* or *fresh vanilla* fragrance. These products are often labeled—the height of irony—*deodorizer*. This could be the ultimate experiment for Maniq participants.

These words are not linked to the materials they emanate from, or to the other senses. They are specific and abstract terms that exclusively describe the olfactory universe.

Objects associated with an olfactory word in the Maniq culture have little resemblance to those found in Western cultures. So these words cannot be translated. The same smell can refer to edible and inedible things, plants and animals, individual objects, activities and locations. For the Maniq, the sun is like the center of olfactory space, with very different projections depending on whether it is a red-hot sun or a white sun. These associations seem strange to us, but they cease to be when we think about perceptual realms that we designate with their own words, such as colors. When we read picture books to our kids, we invite them to point out red, blue, or yellow objects. These objects, of course, take the most varied forms: trucks, people, food, abstract figures. Color becomes its own entity, separate from the objects themselves. It turns out that a large number of cultures coin their own abstract olfactory terms, which define categories that are not recognizable or—here's the key—perceptible to cultures whose lexicons don't include them. Words communicate and shape our experience. When they disappear, even our most intense perception becomes confused and disorganized.

The categories that define us

Categories and words offer a double dimension: they are powerfully good, but carry a stigma. The advantages of using them are obvious. Names allow us to know what we are talking about and compare different sensations, both those we experience and those experienced by others. In a few words, we can express an idea that comes to life in someone else's mind. Someone on the other side of the world, on a rainy Monday, can relate to my sadness on some sunny Thursday. Words allow us to describe an emotion to others at a different time, sometimes centuries later. What's more, they allow us to label an emotion as an entity, as we saw with the languages that have words for smells.

Category building also comes at a cost. Projecting the infinite detail of a continuum of sounds into a few categories means we lose resolution. Our ability to understand that an /a/ is an /a/ when pronounced differently by two people means we are hearing those two sounds as the same. It diminishes our capacity to perceive the difference between the two sounds. It's like seeing the world through a filter that granulates the image into a few pixels. We live in a Minecraft landscape. When we are unable to recognize sounds that other languages have, it's because they are stuck in the same pixel of the auditory space, in the same parcel. This is the advantage and cost of categories: abstraction is gained, but resolution is lost.

This occurs in all realms of thought, from the examples we've already seen—such as sounds, phonemes, and smells—to the one we will delve into later: the emotional realm. Because, just as we are unable to distinguish certain sounds of a foreign language, sometimes we also confuse different emotions that are fused into the same word.

Moreover, words are reflexive in the case of emotions as well. Once we've defined an emotion as *sadness* or *fear*, the complexity we were intending to describe vanishes, and we lose the richness of a world that has infinite gradations, unlike the world of words. As Chesterton points out: "[Man] knows that there are in the soul tints more bewildering, more numberless, and more nameless than the colours of an autumn forest; he knows that there are abroad in the world and doing strange and terrible service in it crimes that have never been condemned and virtues that have never been christened. Yet he seriously believes that these things can every one of them, in all their tones and semi-tones, in all their blends and unions, be accurately represented by an arbitrary system of grunts and squeals. He believes that an ordinary civilized stockbroker can really produce out of his own inside noises which denote all the mysteries of memory and all the agonies of desire."[22]

I found that quote in "The Analytical Language of John Wilkins," the essay in which Jorge Luis Borges analyzes the possibility of creating a language capable of encompassing and organizing all

---

22 When Borges transcribes these words from Chesterton he quotes an essay about G. F. Watts, a fabulous artist who created memorable paintings and sculptures but never wrote a book.

thoughts. I will delve into the details of this text because I believe that in its essence it is, as Borges described it, "perhaps the most lucid thing that has ever been written about language." In the introduction, Borges praises Wilkins's monumental feat: "In the universal language that Wilkins devised in the mid-seventeenth century, each word defines itself. Descartes, in a letter dated November 1629, had already noticed that, by means of the decimal system of numbering, we can learn in just a single day how to name all the numbers up to infinity and how to write them in a new language namely that of numbers; he also proposed the creation of an analogous, general language that would organize and encompass all human thought. John Wilkins, around 1664, undertook that enterprise."

That passage mentions the idea of a mental line on which all numbers are located in simple geometry: we know that 17 is to the left of 34, and 127 is between 100 and 150. This same exercise can be extrapolated to all concepts. Let's imagine that space now as a giant cloud of dots, each of which represents a word. One of those dots, somewhere in the cloud, will be the one that represents the concept *tomato*. Another, in another area within that space, represents the concept *cold*. Once such dots are located, it will be easier to locate other concepts, such as *apple*, which will be somewhere near *tomato*, or *snow* and *ice cream*, which will be in the vicinity of *cold*. These coordinates are not enough, however, to locate most concepts. We don't know if *hope* is to the left of *tomato*, to the right of *cold*, or closer to one than the other. Wilkins's feat can be thought of as an effort to find the geometry of all concepts. Borges describes it this way: "[Wilkins] divided the universe into forty categories, subdivisible in turn into species. He assigned each genus a two-letter monosyllable; to each difference, a consonant; to each species, a vowel. For example: *de* means element; *deb*, the first of the elements, fire; *deba*, a portion of the element of fire, a flame."

In sketching this idea, Borges identifies an essential problem: what are the categories? Just as each language breaks down the space of vocal utterances into its own set of phonemes, Wilkins's categories have nothing essential, nothing unique or particular. Of course, there is no fundamental reason to make *fire* the first of the categories.

Borges explains it with irony and clarity: "After defining Wilkins's procedure, one must examine a problem that is impossible or difficult to postpone: the meaning of the fortieth table, on which the language is based. Consider the eighth category, that of stones. Wilkins divides them into the following classifications: ordinary (flint, gravel, slate); intermediate (marble, amber, coral); precious (pearl, opal); transparent (amethyst, sapphire); and insoluble (coal, clay, and arsenic). [. . .] The whale appears in the sixteenth category: it is a viviparous, oblong fish. These ambiguities, redundancies, and shortcomings recall those that Dr. Franz Kuhn attributed to a certain Chinese encyclopedia entitled *Celestial Emporium of Benevolent Knowledge*. On its remote pages it is written that animals are divided into (a) those belonging to the Emperor, (b) embalmed ones, (c) those that are trained, (d) piglets, (e) mermaids, (f) fabulous ones, (g) stray dogs, (h) those that are included in this classification, (i) those that shake like crazy, (j) innumerable ones, (k) those drawn with a very fine camel's hair brush, (l) et cetera, (m) those that have just broken a vase, (n) those that resemble flies from a distance. [. . .] Notoriously, there is no classification of the universe that is not arbitrary and conjectural. The reason is very simple: we don't know what the universe is."

I have always been moved by this text because, beyond its formidable lucidity to identify the roots of thought (the linguist Steven Pinker and the philosopher Umberto Eco, among many others, have used it as a mainstay of their essays), it is an ode to the human. Without making it explicit, Borges moves between a jocular and admiring tenderness toward John Wilkins and, through him, toward our irrepressible exploratory vocation. In his odyssey, Wilkins comes up against elementary errors, masterfully parodied in the ineffable Chinese encyclopedia. In his coda, Borges distinguishes the essential from the particular and, at the same time, defines science: "The impossibility of penetrating the divine scheme of the universe cannot, however, dissuade us from planning human schemes, even though we are aware that they are provisional. Wilkins's analytical language is not the least admirable of those schemes. It is composed of contradictory and vague classes and species; its device of using the letters of the words to indicate subdivisions and divisions is undoubtedly ingenious. The word *salmon* tells us nothing; *zana*, the corresponding word, defines (for the man versed in the forty categories and in the classes of those categories) a scaly river fish with reddish flesh."

The periodic table of emotions

Can a language like Wilkins's be designed to describe the universe of passions? Is there a set of fundamental emotions that can be recombined to describe all the ones we experience? Could this classification somehow be universal? And if so, how many fundamental emotions are there? Four, six, twenty-seven? It is a debate as old as it is contemporary. Authors belonging to various currents of thought—from Aristotle, Thomas Aquinas or Descartes to William James[23]—have reflected on the passions and, in so doing, have asked themselves again and again about the existence—and the possible number—of fundamental emotions.

An emotion is universal when it is inherent to the human condition and therefore observed in all cultures regardless of the educational tradition of each society. This assumes that it must also have traces in early childhood. Going back even further, an emotion is universal if it is part of our genetic legacy and must, therefore, have precursors in nearby species.

There is no consensus on the existence of fundamental emotions that meet all of these conditions. The first systematic journey in search of a universal origin of emotions was undertaken by Charles Darwin. After achieving great renown with the publication of *On the Origin of Species*, Darwin collected data on emotional expressions from different parts of the world, from his immediate surroundings to the most remote corners of the globe. His search had begun much earlier. On his famous voyage aboard the *Beagle*, he asked to be

23 This list is a humble tribute to the mythical football match between Greek and German philosophers dreamed up by Monty Python. The German team's line-up included Leibniz, Kant, Hegel, Schopenhauer, Schelling, Jaspers, Schlegel, Wittgenstein, Nietzsche, Heidegger, and . . . Beckenbauer. Marx was a substitute.

informed of all the emotional expressions observed in the southernmost corner of the world: Tierra del Fuego. He was also interested in the facial expressions of newborns and collected early data minutes after the birth of his first son, William Erasmus. During his son's first days he noted sneezing, hiccups, yawns, stretches, screams, and, above all, tickles. He did the same with each of his ten children and then combined these observations with information he carefully solicited from people around the world who were well placed to observe babies, the blind, the insane, and a highly varied range of the human race. He did the same, of course, in the world of animals. He collected information from his pets and from his visits to zoos, and through the eyes of naturalists and elephant keepers he beleaguered with his questions. He concluded that emotional expressions fulfilled an adaptive function and that they resulted from an evolutionary process that humans shared with animals.

More recently, Paul Ekman, a professor emeritus at the University of San Francisco, became one of the most influential scientists in the world by declaring that this periodic table of emotions exists:

that there are universal expressions of anger, sadness, fear, and happiness, and that humans and certain animals can produce and recognize all of them, even babies, and people from the spectrum of different cultures can recognize them easily, almost automatically. Scientists such as Lisa Feldman Barrett have questioned this idea, with examples that indicate emotions are much more mutable than we intuit, like phonemes in a baby's universe or spatial representations of time in different cultures.

Without a doubt, facial expressions of emotions are not as unambiguous as Paul Ekman suggested. Lisa Feldman Barrett clearly exemplifies this ambiguity with the image of tennis player Serena Williams after winning a Grand Slam title. Her face, which seems to express anger and hatred, is actually one of sublime celebration.[24] Yet perhaps the space of emotions is also not as flexible and unstructured as Feldman suggests. The statistical observation that many cultures have a word to refer to the same emotion, such as sadness, reflects an ordering tendency in the tabulation of emotions. In the same way, the existence of emotional precursors that we share with an entire fauna of living beings—such as responses to pain, cold, and hunger—suggests a genetic anchorage for many of our emotions' central functions.

This battle around universality is not exclusive to emotional territory. There have been very similar debates in many realms of human cognition, between those who believe that human faculties—such as language—are innate and biologically determined, and those who believe they are primarily cultural constructions. These discussions usually play out in a dead heat, with discernible signs that the expressions of human behavior emerge from a biological fabric that shapes and outlines a space of possibilities. In turn, this space is vast and accepts a wide range of inputs and influence from the culture, education, and unique life trajectory of each person.

The search for the balance point between rules and freedoms is advantageous from a purely pragmatic perspective, beyond any

24 There is a homemade version of this experiment. All we need is a mirror and an orgasm.

philosophical discussion of the nature of emotions. Understanding that we have leeway to change the contours of our emotional experience is a good starting point for improving it. But so is understanding that sometimes this process causes friction because it rubs against basic elements. In short:

1. We inevitably reduce the accumulation of our emotional experiences to a few words that form the basis of our emotional language.
2. These categories are not universal, nor completely arbitrary. The example of phonemes again comes in handy: each language has some freedom when choosing how to divide sounds into vowels, but that division also has some regularities resulting from what we can pronounce and, thus, our brain is tuned to hear. So, while there is a wide diversity of vowels in the world's languages, they all include the /a/, the /i/, and /u/, precisely because these three are the easiest to pronounce and hear. The same goes for time and space; while there are whims of language to draw projections of the past and the future, they coincide in their organization along a line (be it straight or circular) due to the (uni)dimensionality of time.
3. The correlation between the continuum of emotional experiences and the words described by each emotional category can be ambiguous, as when we prepare our mouths to pronounce one vowel and say another.
4. The choice of a word (a category) to designate an emotional experience is reflexive. Remember the analogy about belief in the financial market that we saw earlier? The multitudes within us can drive stock prices up. There are real facts (an illness, a person we know, a death, a fight, a kiss) that change each emotion's *share price*, or value. Occasionally there are also financial bubbles, illusions of the system that shoot up value and are perpetuated reflexively. The categories we choose to describe emotions determine and

condition our conscious experience of them, their impact on our bodies, and the things we do as a result, such as screaming, laughing, crying, insulting, or hugging.

Medieval passions

With these principles in mind, let's briefly review the intuitions great thinkers have had about the geometry of the space of passions, beginning with the sixteenth century. In *The Passions of the Soul*, Descartes proposes a list of the primary emotions that correspond with most descriptions: wonder, love, hate, desire, joy, and sadness.

Some four hundred years earlier, Thomas Aquinas proposed eleven fundamental emotions. This was not a Kabbalistic or random number, but the result of precise mapping. For Saint Thomas, emotions—which he called *passions*—were part of the broader principles that set into motion all bodies, living or not. In his conception—later also in Freud's and many other theorists of human psychology—the image of the force that compels bodies into action is more than a foundational metaphor for understanding desire. The etymology of *emotion* is eloquent. It sets us in motion.

Just as we often borrow from the forces of physics to explain emotions, the verbs of passion are turned into metaphors for the dynamics of inanimate matter. To describe gravity, Thomas Aquinas uses the figure of a heavy body that *wants* to fall to the earth's surface, where it *enjoys* rest. Being at the center of the universe is good for the stone, not in the moral sense, but as a teleological purpose that sets everything in motion.

In this concept of dynamics, all things tend toward good, which today we would call *balance*. The difference between a stone and a living being (animals, for Aristotle and Aquinas, have passions) is that the good moves the stone even when the stone does not realize it. However, living beings have to perceive the good to feel that attraction. A dog chases prey only when it perceives it. Passion is the drive of a living being towards good, towards a conclusion, towards balance.

In Aquinas's diagram, the passions are ordered according to the immediacy of the present (eating or drinking) or a future perspective (caring for a baby, studying). The former he calls the *concupiscible*, the latter the *irascible*. The six concupiscible passions come in pairs: love and hate, desire and aversion, and joy and sadness. Not all passions have the same rank. Love and hate are the first passions and cause all the others. Then there are the five irascible passions: hope and despair, confidence and fear, and the passion that breaks the parity: anger. It does not fall into the category of good or evil, but swings between one and the other. It is the force that allows us to move from fear to audacity.

An idea appears in the writings of Saint Thomas that still today remains a vital concept in emotional regulation: anger is usually an emotion harnessed to turn fear into audacity, like when athletes summon their fury to win a match. In a way, this is at the core of almost the entire plot of *Star Wars*: fear leads to anger, anger to hatred, and hatred to suffering.

The various taxonomies not only differ in their identification of primary emotions, but also in their tracing of their directions through space. For example, a twentieth-century model developed by James A. Russell is based on three-dimensional scales: one that measures pleasure; one that measures arousal; and one that measures dominance. The first axis is the simplest: joy is pleasant and fear is unpleasant. Arousal, for example, makes it possible to separate

boredom from anger: both emotions are unpleasant, but anger is highly arousing while boredom is not. Similarly, two unpleasant emotions such as anger and fear are separated along the third axis: anger is dominant and fear is submissive. Russell presupposes in his theory that these three axes uniquely describe all emotions. Each emotion, including the most complex ones, can be characterized according to its value along the axes of pleasure, arousal, and dominance.

The wheel of emotions

Our final stop on this brief review of the history of emotional geometry is at the famous wheel, conceived and diagrammed by the psychologist Robert Plutchik.

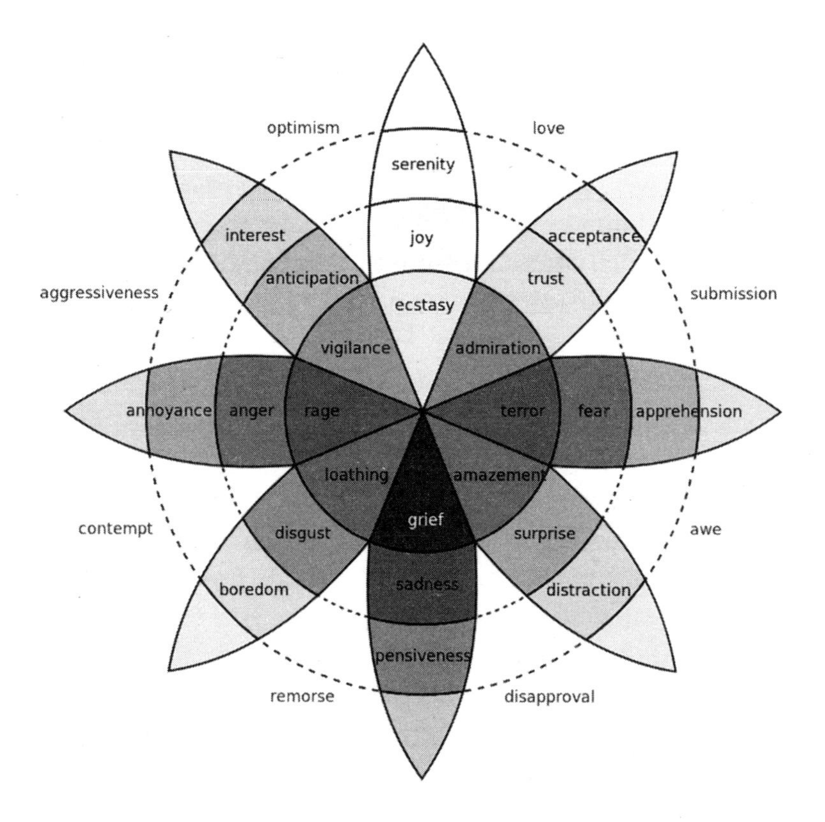

He introduced an innovative idea: the emotional space he mapped is circular, like the Aztecs' conception of time. This allows you to travel from one emotion to another by turning the wheel in either direction. For example, we can go from sadness to anger along the following path: from sadness to grief, and from there to loathing, which becomes rage and triggers anger. This is the most direct path but not the only one; we can also travel, in a much longer chain of associations, in the opposite direction.

This brief review reveals similarities and associations that can be seen as a common point in the controversy between Feldman Barrett and Ekman over the existence of universal emotions. Ekman's argument is that there are some emotions that are repeated in all classifications and that have consistent expression among individuals from all cultures: sadness, anger, joy. Feldman Barrett's argument is that there is nothing fundamental about these emotions, since we can reappraise them at will. Serena Williams's expression, which at first seemed like rage, becomes ecstasy when placed in the right context. Barrett extends this principle to all other emotions, arguing that they can be exchanged for others depending on context, culture, education, and interpretation.

The intermediary territory I want to describe is an organization of emotional space into *fuzzy categories*. The lack of defined boundaries gives us great latitude for reinterpreting the emotional experience. At the same time, the emotional experience is linked to a geometry—like Plutchik's wheel or Russell's axes—and therefore some reinterpretations are simpler than others. Let's make an obvious analogy: you can mistake a tangerine for an orange, but you would be unlikely to confuse a plum with a sardine. The same goes for phonemes: confusion between /r/ and /l/ occurs more frequently in some cultures than between /t/ and /p/. In each of these realms, the elements that get confused are the ones that are "closest" to each other in the complex representational space of the brain. In Serena's example, in fact, that confusion is revealing. Ecstasy and rage are two highly arousing emotions—leading them to overlap along one of Russell's main axes—and this makes them more likely to be confused.

The architecture of hell

The selection of cases in our very brief historical review is capricious, but it works to illustrate how our view of emotions evolved. We began with a mere enumeration, which became a hierarchical organization where emotional space is structured into categories and subcategories. This organization grew more sophisticated over time, and in the previous century there was a fundamental shift: emotions began to be represented in geometric space, allowing us to establish their distance from each other.

Russell's space has three dimensions; Plutchik's wheel, on the other hand, is on one flat plane. This makes it clearer, because that is precisely where we can draw—literally illustrate—our ideas. The plane is also the space of art, the pictorial space in which passions have always been the object of representation. A classic case is the triptych by Bosch (circa 1510), which depicts paradise on the left and hell on the right, panels that function to frame the wide central panel: *The Garden of Earthly Delights*. The scene is confusing, with no hierarchies. There is no definitive agreement on what the painting represents, whether it depicts passions in the world or a sort of paradise without original sin. The eye travels around the panels: there is a swimming pool, a fountain, and women surrounded by young men riding animals. There are details that are difficult to interpret, some elements out of scale and others that defy the laws of gravity. As in classical thought, *The Garden of Earthly Delights* is a description of the passions in no apparent order.

This order begins to emerge in medieval literature, with *The Divine Comedy* and its architecture of hell. Even the chronology of Dante Alighieri's journey imposes a hierarchy. This schema offers a first grouping of sins as excessive (gluttony, lust, and greed), deficient (sloth) and misdirected (envy, pride, and wrath).

Dante's speculation about the order of passions, like those of the other great thinkers and artists of the past, is subjective. Is it possible to construct a map of passions that, instead of representing an enlightened opinion, incorporates the intuitions of an entire collective? In order to answer these questions, we enlisted the help of science.

We asked a large group of people to imagine their own map of hell, placing each of the seven deadly sins within a white square. If there were no correlations in how each person organizes the cardinal sins, the result of superimposing all these drawings would be a big amorphous blob, but that wasn't what happens. When we combined the thousands of individual drawings, a coherent structure emerged: at the bottom, wrath, greed, and pride are piled up, envy is in the middle, and, finally, in a higher triangle, are lust, sloth, and gluttony. It represents a collective architecture of hell.

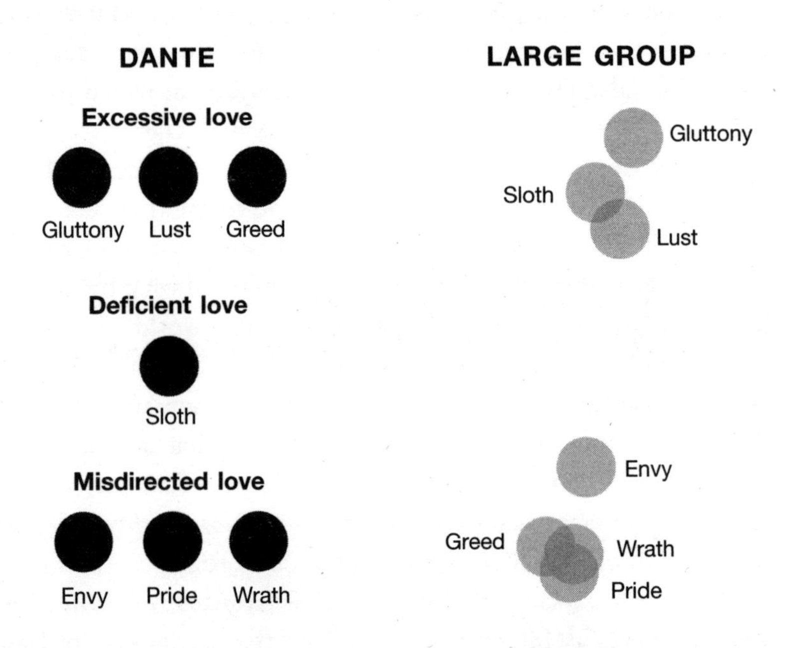

This mental architecture of sins seems reasonable. In one corner are the hedonistic vices, food, rest, and sex galore. In the other are the sins most associated with suffering and evil—such as pride or wrath—that are not linked to pleasure. In the middle we find envy, which is both desire and suffering. This order resembles Dante's, but has some significant differences. Dante placed sloth opposite lust, while the collective mind situates it in the same neighborhood as gluttony and lust.

These constructions are structures that allow us to organize and assign meaning to the passions, just as the memory palace does with memories. In a way, fiction and philosophy have given us many palaces, each with its own logic and relationships, which enable us to order difficult and dull aspects of the passions.

The paths of words

The idea of a geometry for the passions, with its notions of vicinity and adjacency, illustrates a broader principle that regulates all thought: concepts are combined into a network of relationships by proximity or similarity. In other words, they are defined, above all, by the place they occupy in the network of meaning. An example is the concept of *single*. What does it mean? Not *married*. And what is *married*? Having *wed*, which means, in turn, joining in matrimony to another person through legal means . . . Defining a word requires other words; their meanings are intricately tied to others, as in the dictionary, in a loop that can go on for ever. This recursive capacity of meaning, as Charles Peirce pointed out more than a century ago, is at the heart of language.

As with people, each word is defined, to a large extent, by the neighborhood it inhabits. Tell me who your friends are and I'll tell you who you are. This idea, which would later be revisited by Quine, Wittgenstein, and many other philosophers of language, is of enormous importance for our purposes. Since it is quite complex and abstract, we will break it down from different perspectives.

Let us begin with the work of Thomas Landauer and Susan Dumais, who applied the network of concepts to artificial intelligence research and offered a modern solution to Plato's problem, which has tormented philosophers and psychologists for centuries: how is it possible that we all know so much more than we've been taught?

The examples of this paradox are multiple and incongruent: a child refuses to eat some rotten food even though he has never tasted anything like it and no one has taught him that spoiled food can make you sick. Beyond this type of knowledge, which we call *instinctive*, the paradox is also applicable to more *cognitive* spheres. For example, the linguist Noam Chomsky pointed out that the phrases we hear in childhood do not provide enough information for us to infer grammar, yet we acquire it with surprising ease. This allows us to create new and grammatically correct sentences even before learning to calculate three plus four. We see this again in vocabulary acquisition. In Western societies with mandatory schooling, a ten-year-old child incorporates about ten to fifteen new words per day, inflating their lexicon to the approximately thirty thousand that the average adult English-speaker knows.[25] It's a dizzying pace. Yet what is most extraordinary about this learning

25 As if that weren't enough, there are those infuriating people who pepper their speech with big words they don't know the meaning of because they think it makes them look more hypotenuse.

curve is that children are not taught more than three words a day at home or school. How do they discover the rest? The evidence is overwhelming: every day children learn the meaning of many words that no one has taught them.

The solution Plato came up with for this paradox is that we are already born with all knowledge and throughout our lives we gradually remember it. Centuries later, a psychology professor, Thomas Landauer, and a computing professor, Susan Dumais, offered a very different solution: every time our brain (or its algorithm) finds a new word, it makes a conjecture that the word's meaning is in the center of the cloud formed in the space of meanings by the words in the same sentence or paragraph. This simple and effective idea laid one of the foundation stones for modern cognitive science, computer theory, and artificial intelligence: Quine meets Google and Facebook.

Let's see how it works. Our irrepressible propensity for induction leads us to establish relationships in order to predict the unknown from scant data. Let's look at an example: we have a bag with millions of balls inside it. We pull one out at random and it's red; we pull out a second one and it's also red; we take out a third, a fourth, and a fifth, and they're all red. What color will the next one be? Red, of course. What is obscured by the salience of the five repetitions is that there are millions of balls in the bag, about which we know absolutely nothing. The example connects induction with availability bias, which, as we have seen, leads us to make bad decisions by convincing us of a hypothesis that only makes sense based on the small amount of evidence we are able to consider.

Induction is also a creative tool, both a help and a hindrance, as we saw with the RAT test for the *house, family, apple* series. *Tree* is an acceptable solution because it establishes a semantic link with each element of the series. From a geometric perspective, in the space of meanings, this is equivalent to saying that the word *tree* is located in the center of those three points. Verbal creativity is also based on inductions made by navigating the space of meanings. Just as this algorithm allows us to search for a word located in the center of other known words, it can also be used to discover new words whose definitions we do not know. This same idea can be implemented to understand how, as children, our web of vocabulary meanings grows and evolves.

The algorithm that Landauer and Dumais implemented is based on the simple assumption that semantically related words usually appear close to each other in texts. The algorithm *reads* an extensive body of text, let's say all the works written in Spanish in the past five hundred years. Every time a new word appears, it is incorporated into the network according to its relationship with neighboring words. This is how we learn: by incorporating each new element into its most likely neighborhood based on the context in which it appears. We can feel the algorithm at work inside of us, when we find a new word while reading and intuit its meaning from the place it occupies in the text. It's not much different from what happens in social networks, both virtual ones and those we build with our friends and colleagues. When we meet someone for the first time, we tacitly infer a myriad of traits about them based on their clothes or way of speaking. If that someone is not alone, their companions will give us more clues, as happens in social networks with followers and friends in common. Knowing their profession and schools attended will allow us to make many more inferences. These are the threads that link that person to different attributes. We welcome new words into our vocabulary in the same way.

The most extraordinary thing about Landauer and Dumais's work is that they went beyond the theoretical plane and translated their idea into an algorithm that, by the end of the last century, worked on a rudimentary computer, processing texts just as a child would: every time it came up with a new word, it inferred the word's meaning from its proximity to contiguous words. This algorithm was one of the first great successes of what we now call *artificial intelligence*. It could find synonyms, infer meaning from words it had never encountered before, and was able to pass an English test required to enter an American university. This was the evidence that solved Plato's problem: a relatively simple computer program that implemented the principle of induction using co-occurrence was capable of learning thousands of concepts without a teacher, simply by reading.[26]

Words and emotions

The principle of induction allows us to use conjecture to create new concepts. We can imagine a line within the semantic neighborhood of "animals" that extends from the concept *cat* to the concept *tiger*. Along the way there will be other felines, such as the puma. If we extend that line further, we can create a caricature, an imaginary animal whose features are an exaggeration or deformation of the traits that distinguish a tiger from a cat. We can also invent characters by fusing more distant points in the space of meanings. This can be seen in the history of fables, in catalogs of medieval fauna, and in cathedral bestiaries, which feature hybrid monsters formed by a mixture of recognizable parts: the Chimera—with its goat's body, snake's tail, and lion's head—is finally defeated by Bellerophon on the back of Pegasus, a winged horse. Dragons have an even more complex structure. They are creatures that breathe fire, and have the body of a snake, wings (of a bat, perhaps), and the head of . . . a dragon.[27]

Armed with this new inductive tool, let us now return to Plutchik's wheel to *create* emotions by induction and recombination. For example, the combination of *joy* and *acceptance* produces a particular version of love; the overlap of *anger* and *disgust*, a nuance of hostility. Plutchik added colors to his wheel precisely because he thought of emotions as capable of being combined in a palette, like primary colors that make new hues. This game can be played recursively to

26 The use of verbs such as *read*, *explain*, and *learn* applied to a program designed to run an algorithm on a computer might seem like a joke, but is it? All humor shares an element of the unexpected. In this case, it is the unexpected proximity of *read* and *computer*, words that once were far away from each other in the network of meanings. However, the network is evolving and those words are getting closer. Maybe it once was a joke, but it's not any more.

27 We can also trace the path backwards: what animals is Totoro made from? And Stitch?

increase the granularity or specificity of an emotion: what lies between sadness and melancholy? Is there a word that designates such a specific emotion? Maybe we should mix emotions with concepts from other neighborhoods. Let's take for example "the mixture of sadness and agitation of a Sunday afternoon spent stewing over Messi's missed free kick, as if somehow you could bend it two millimeters so it would enter the goal, allowing Lionel to hoist up the cup, and you to finally get some sleep." The more we specifically refine the process, the more evident its social and cultural baggage becomes.

In another context, in Norway, there are months of darkness in the winter during which you long to drink a beer in the sun. Norwegians famously have their own word, *utepils*—which literally means "to drink beer outdoors"—to designate that particular desire. Each culture, each community and each person has its own palette. In his book *The Dictionary of Obscure Sorrows,* John Koenig lists a series of emotions that, like *utepils,* only have specific words in some languages. In Mandarin *yù yī* is the longing to feel again with the intensity of a child; in Polish, *jouska* is a hypothetical and compulsive conversation that takes place in our mind; in German, *Zielschmerz* names the fear of getting what one desires. Are those emotions less real once we learn that Koenig made the words up?[28]

The discovery of an emotion

The *game* I propose entails describing a feeling that is especially meaningful to us, and using this description as the definition of a new word. Why is it important to have a specific word to label an experience? At the end of the day, the virtue of language is its ability to recombine words in order to express any concept. It turns out that, by giving an experience its own name, we encapsulate it; we create a succinct, precise, and stable way of talking about it. Think

28 There are some real examples. *Schadenfreude* in German is the unmistakable feeling of joy caused by the misfortune of others. In the US and Mexico, *the munchies* refers to the uncontrollable hunger after smoking cannabis. The *guayabo puntudo,* in Colombia, describes sexual arousal on a hungover morning.

of the word *meme*. We all know what it is. However, if that word didn't exist, it would be tiresome and complex to describe that concept.

Discovering and creating new words is one of the most effective ways to take the helm of our emotional experiences. They can be words that we do not know from our own language, such as *twitter-pated*;[29] words that come from other languages, such as *utepils*, or that we've invented, like *jouska*. It doesn't matter. The important thing is finding a useful mixture of colors on the palette of emotions and coining a single word we can use whenever we want, without getting lost along the way in a lengthy, cumbersome description that varies each time. These new words work like a magnifying glass to identify and express what is happening with us, or like a compass to lead us to interesting places in our emotional life.

29 "Infatuated or obsessed; in a state of nervous excitement."

Conversely, a word can be harmful when it becomes a huge sack where different emotions get jumbled up. This is what happens, for example, with the word *love,* which we fall back on to express a wide range of feelings, such as the bond that unites us to a child, a friend, or a partner. And within relationships, we can even use it to refer to both the ardent feeling of the first days and the serene affection that builds over years of sharing a life. This, of course, causes all kinds of confusion. When someone says, "I'm not in love any more," what they may actually be wanting to imply is that they feel another kind of love. Confusing this mutation for a loss, because we lack more precise terms to describe what we feel, can lead to unnecessary disappointment.

Some people discover caves, planets, theorems, insects, or rivers. About fifty years ago, Michelle and Renato Rosaldo, on the Philippine island of Luzon, discovered an emotion. The Rosaldos are among the few anthropologists who have lived on this island with the Bugkalot natives. This is explained by the enormous difficulties of the trip, by the hardships of cross-cultural communication, and because the Bugkalots used to be headhunters. The Rosaldos studied the emotions of this culture and they found that all but one had correspondence with the emotions of the West: the one that the Bugkalot call *liget,* which is, for Westerners, almost as incomprehensible as the olfactory vocabulary of the Maniq.

Renato Rosaldo first came into contact with this emotion one day when he found a Bugkalot man brimming with such physical energy that he couldn't stop cutting down trees while shouting, "I have *liget*!" From his Western perspective, Renato associated the word with an emotion of irrepressible joy. Later, he realized that the trigger for *liget* is usually someone's death. Curiously, this grief is not expressed through tears, but by running rampantly to cut down trees or cut off heads, amid euphoric songs. This intersection of triggers, effects, forms of expression, channels of relief, and communication makes that emotion extremely difficult to translate, and even more difficult to feel.

Renato says that his first personal experience with *liget* happened in 1981. Returning from a walk he felt a terrifying silence, as if the entire village had suddenly gone silent. His wife Michelle had fallen off a cliff and, as he approached, he saw her body lying already lifeless. There he felt an overflowing energy, as if he and the whole world were spinning, expanding and contracting. Some time later, while driving on the road back in California, he began to feel an uncontrollable anxiety, a pressure he couldn't bear. He stopped on the road, got out of the car and started howling. He immediately knew what was happening to him: it was *liget*. Only then did he find the most accurate words: high voltage. What he felt was a kind of frenzy, of deep pain, of outburst; like a bolt of lightning running through him. After years of studying that emotion as an anthropologist, Renato's body was prepared to experience that complex mixture of sensations and manifestations, that strange combination of vocalizing, relief, and bursts of uncontrolled energy.

The nature of emotions

The story about *liget* illustrates both the most extraordinary and the most problematic aspects of studying emotions. What is extraordinary is the elasticity of mental life, our amazing capacity to transform our emotional experience even in the most difficult territory: grief. What is problematic is that this sensational power is trapped in a muddy bog of vagueness. Despite everything we have learned and discovered, it is still very difficult for us to agree on what an emotion is. Many discussions about the nature of emotions are ultimately mere semantic distinctions.

A year ago I offered a course at the Instituto Baikal with Christián Carman, Sergio Feferovich, and Diego Golombek. It sounds like the set-up of a joke: a philosopher, a musician, a biologist, and a neuroscientist teach a course . . . We are good friends and we were feeling open to learning from each other, so we created a suitable conversation space.

In the first class I talked about the categorical paradigm shift that occurred when some European glass makers built the telescopes with which Galileo first glimpsed something impossible to see with the naked eye: satellites orbiting Jupiter and Earth. That allowed Galileo to irrevocably change our conception of the universe. Then

I talked about dreams, which are so elusive that it seems impossible to determine whether they are just deceptive memories created by the brain when we awaken. Or rather, it seemed impossible, because now we have instruments that allow us to see brain activity, which were used by the Japanese scientist Yukiyasu Kamitani to do something that, again, up until then was impossible: reconstruct, in real time, a dream's plot based on the brain activity of the dreamer.

Is there anything we can now see, in what was once invisible, that changes our view of emotions, a view carved out over centuries of philosophy and introspection?

Intuition tells us that emotions are, first and foremost, a mental experience. We feel sadness, joy, love, anger. But this is just the tip of the iceberg of a much more intricate phenomenon. When we observe emotions with the naked eye, it is much like how the heavens were observed prior to the Renaissance, but in reality they are much richer, involving a complex repertoire of bodily and cerebral responses.

Bodily expressions are so fundamental to emotions that, just by mimicking them, we can induce an emotional experience. As a kind of epistemological exercise, or thought experiment, I suggested in that course that we take this idea to its limits. What if, at the end of the day, the conscious experience of an emotion is almost a mirage? Can an emotion be completely unconscious?

As strange as this idea may seem, it is actually commonplace. When we say that a mouse—or a baby, for that matter—is afraid, what we are saying is that we have observed it producing a series of bodily changes, expressions, and reactions that we associate with human fear. To that we then add the whole mass of brain processes and pharmacological modulations that we know from scientific research. We know all of those things when we say that a mouse is showing fear. The only thing we don't know is whether it is actually afraid.

I want to show that it is possible to induce unconscious emotions, first in a shallow sense and then more profoundly. The shallow sense is very easy to demonstrate. There are countless experiments in psychology that show that almost every aspect of behavior, both

cognitive and emotional, can be conditioned by stimuli of which we have no conscious memory. A classic experiment is to quickly flash a number on the screen for a fraction of a second so small that it becomes invisible. Then, ask the person looking at the screen to choose a number, any number, and they will most likely opt for the one they "didn't see." Although the visual experience was not conscious, their brain registered the number.

This experiment has also been conducted in the realm of emotions. Subliminal exposure to a sad face induces mimicry, and participants respond with a sense of sadness. They do not know what to attribute the overwhelming sensation to, because the stimulus that triggers it is unconscious, eliciting what are called *free-floating* emotions. The participants come up with any number of reasons to make sense of that sadness, just as false memories are created to give coherence to other inconsistencies in life.

The bodily responses to a subliminal image of a sad face are immediate, yet there is a lag before the participants claim to feel that emotion. In that period of time when the participants' whole bodies express sadness while they say they do not feel it, are they, in that limbo, experiencing an emotion? We find that this question is very much like the idea of *fear* in a mouse, or any other animal, or another person, for that matter. We perceive the emotion with the naked eye when their bodies express it, without questioning what they are feeling. This is the *profound* expression of an unconscious emotion.

Christián, the philosopher, had a lucid and succinct response. He said that we'd subtly changed the emotion by burdening it with its accompanying biological trappings. Next, we defined the emotion as its biological trappings, and then showed that all the biological flourishes can exist without the emotion being experienced. Along the way, he said, we forgot that the essence of emotions is studying what happens to us, what we feel, what makes us suffer or love, which is at the root of all these questions.

One could conclude that the physiological dance that overtakes the body is irrelevant when sadness is not actually felt. But it does matter, even from a purely pragmatic perspective. The bodily expression of an emotion can make us sick, or heal us, even if we don't

consciously perceive it, and—beyond our own experience—always ends up affecting others. I've had someone ask me why I'm angry and been surprised at the very question. Me, angry? *It turns out that sometimes our emotions are more visible to others than they are to ourselves.*

The reason behind emotions

Discovering the chorus of physiological and behavioral manifestations of an emotion is one way to understand its reason for being. In the face of danger, fear prepares us for flight, and anger for a fight. This was how Thomas Aquinas viewed emotions, as a reaction to states of tension that seek to restore balance. Robert Plutchik, who created the wheel, suggests that a good way to understand emotions is to expose the primary functions they perform.

The idea is clearer if we give an example: what guarantees our survival is timely flight, not fear. Once we start thinking about emotions this way, as spurred by vital functions, we find examples in most species: the search for food, the response to confrontation in the form of fight or flight, reproduction, self-preservation, and a tendency to explore. The relationship between functions and emotions is not a one-to-one correspondence. The response to confrontation, for example, can be one of fear but also one of anger.

The vital functions are expressed in disparate ways in different species and individuals. For example, flight, mimicry, camouflage, and confrontation are all ways of solving the same problem: surviving a threat. What is universal is having a quick mechanism for responding to a threat, not the mechanism's bodily expression. For this very reason it is difficult to find a universal gesture denoting anger, or any other emotion.

The biological circuits that implement these functions have been observed in all sorts of organisms, even the most primitive, such as viruses, fungi, bacteria, and algae. These mechanisms are so ubiquitous that they are even found in plants, whose response to touch through chemical signals allows them to retract, paralyze, attack, and camouflage. These examples may serve to reconcile us to the idea that there can be fear without any feeling or awareness of it.

The automatisms that are triggered in response to danger can be understood as a precursor to the conscious sensation of fear. That sensation is just one more ingredient in this whole cascade of responses. In the distorted magnifying glass of the naked eye, mental experience is the epicenter of an emotional response. However, in the immense universe of species, it seems to be just one more on a very long list of occurrences.

A less wide and strange world

Different versions of Landauer's algorithm are now routinely used to automatically identify the meaning of a text or conversation. In my research group we have used this tool to infer certain elements of thought, based on language. Just as there are applications that notify us about our physical activity with the calculation of steps and heart rate during a walk, the tools described in this chapter can give us a statistical summary of our mental activity. Embedded in the space of meanings, the words we express offer a privileged window through which to see—and record—fluctuations in our mood, transformations in our thought, our recurrent and obsessive ideas, our joy, our depression. Our words say so much about us.

Some time after the birth of Milo, my first child, I discovered that the word *careful* had become, overwhelmingly and despite my best intentions, the word I most frequently used. In some cases, of course, its use was justified. Be careful when crossing the street is fine; careful with that chair, perhaps; but careful with that plum, or that spoon? My abuse of the term had made it a bane, and my desire to resolve this awkward obsession was one of the primary motivations behind the research that led to this book.

Just as, from time to time, we need to be reminded to go out and move our bodies, we occasionally need to be encouraged to change our ideas. We need someone to call our attention to our misuse of harmful words and invite us to rethink it. Because choosing our words is like choosing what to put on in the morning. The decision to dress in bright or muted colors not only changes how we present to others, but also our mood. The words we use give shape and color to our world, and to the world of the people we love most.

Exercise: Ideas for living better

1. **You can change your emotional experience**
   Despite how it seems, what you feel, how you feel it, and how it affects you is not etched in stone or out of your control. Being aware that you can change these experiences is the ideal starting point for directing your emotional life along the path you choose.

2. **Don't ask the impossible of yourself**
   Demand only those things that are within your possibilities, in your zone of proximal development. Reasonable goals help us to move forward, while unachievable ones can be a source of frustration and loss of motivation. This is not simply a matter of deadlines. In any case, goals can always be revised.

3. **Remember that your language conditions how you feel**
   The words you know and use define the borders of the landscape of what you feel, of how you describe yourself, and how you explain your biography. Reflect on your emotions, their details, connections, and differences; reflect on the words that describe them. Thinking about your emotional repertoire expands your range of options, both in terms of what you feel and how you react. Without words, our perception is muddled, even our perception of very intense experiences.

4. **Discover (or create) new words to describe your emotions**
   Giving a precise name to a mixture of emotions gives us a

clearer picture of what we are feeling and, as a result, more control over how we deal with those feelings. Strive to have a wide palette of emotions, an encyclopedic catalog. Avoid categories that are too broad, as they provoke distortions, misunderstandings, and unnecessary pain and disappointment.

5. **Place particular emphasis on the middle zones**
   We often lack words to describe the intermediate points between two well-defined emotions. What lies between sloth and sadness? What's the word for the feeling between joy and surprise? Seek these words out or create them in order to gain higher resolution and have a less pixelated and more personalized emotional experience.
6. **Your body not only reflects what you feel, but also conditions how you feel**
   The bodily expression of an emotion, despite often being subconscious, also has the power to induce it. Cracking a smile can produce happiness (although fleetingly) and frowning can lead to anger. Being aware of this phenomenon and paying due attention to it allows us to take the reins of an essential part of our emotional experience.
7. **Change your words to change your mood**
   The words we use give shape and color to our world and can alter our emotional experience. A good starting point, if we want to shift our mood, is to choose different words.

CHAPTER 5

Governing Our Emotions

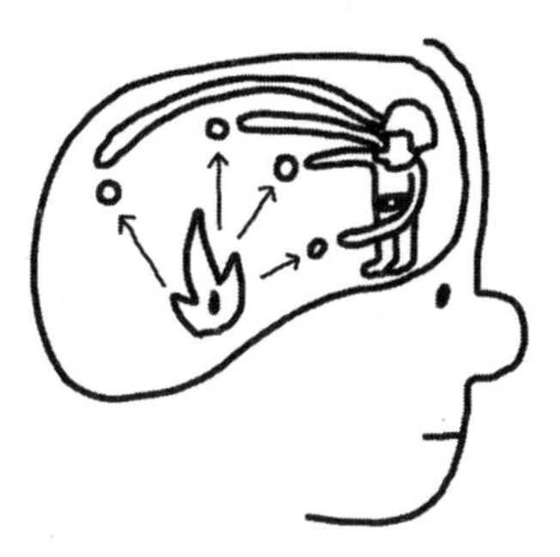

How to take control of our emotional lives

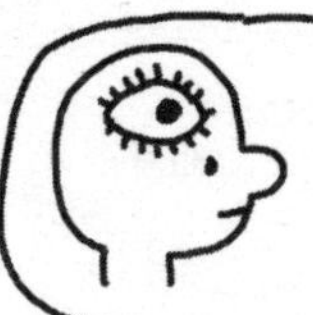

THE COGNITIVE CONTROL SYSTEM GOVERNS WHAT WE THINK AND WHAT WE FEEL.

WHAT WE HEED AND WHAT WE IGNORE

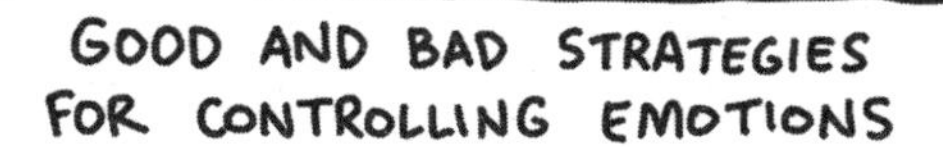

INDUCTION

THE BRAIN DEDUCES FROM OUR BODILY EXPRESSIONS WHAT EMOTION WE'RE FEELING.

SMILING MAKES YOU FEEL GOOD

IT'S NOT THAT HARD TO IMPLANT AN EMOTION

THE EFFECT IS FLEETING AND IMPRECISE

DISTRACTION

LOOK FOR SOMETHING THAT DISTRACTS YOUR MIND.

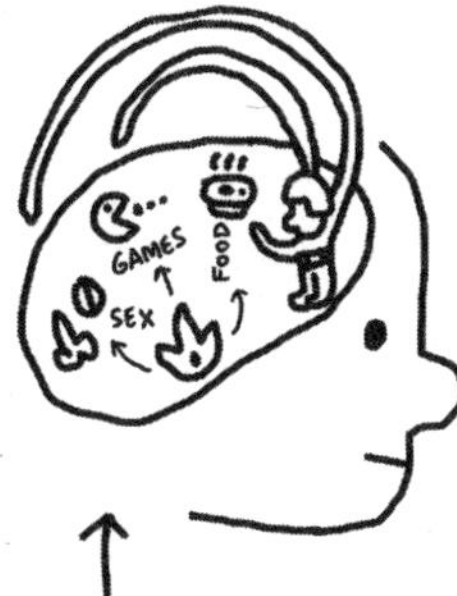

DISTRACTIONS ARE USUALLY ADDICTIVE.
THE CURE CAN BE WORSE THAN THE DISEASE.

...AND THE EMOTION CONTINUES TO DO HARM.

REAPPRAISAL

DON'T SUPPRESS THE EMOTION OR LEARN TO LIVE WITH IT.

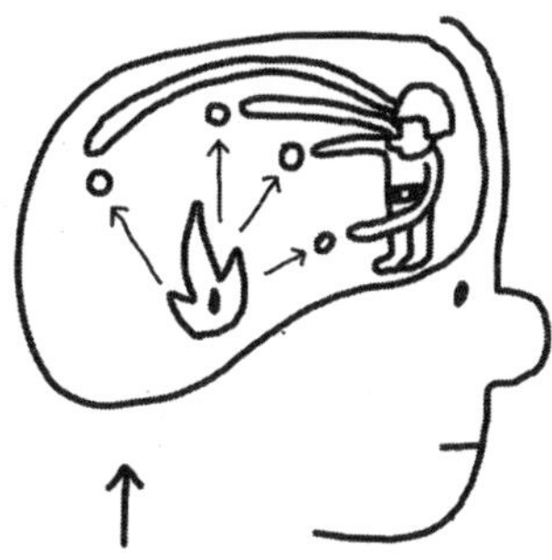

USE LANGUAGE TO GIVE IT ANOTHER INTERPRETATION

FOR EXAMPLE: EMBRACE FEAR.

VERTIGO THAT "CAUSES" PLEASURE

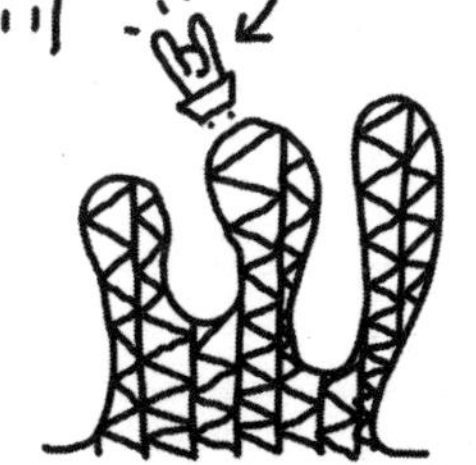

ROLLER COASTER

We have seen how conversation allows us to detect errors in our reasoning and consequently improve our decision-making, and how, contrary to the tendency on social networks, good conversations converge towards reasonableness, even when we are dealing with the most complex issues. We have also looked at how our life story becomes tangled up into a ball that creates our identity. Language is the chisel we use to sculpt our autobiographical story, along with a good dose of fiction. Therein lies its power and its stigma. A deep dive into the atoms of conversation has revealed a more general principle: words form a matrix that organizes ideas in areas as diverse as time, sound, and intense feelings. We will now follow the example of the ancient Greek philosophers, by thinking about the best way to use the power of words to take control of our emotions and, thus, "live better."

We will begin this journey by studying the power of conversations in a very different way: what happens when they disappear? This is the worst deprivation; true loneliness is having no one to talk to. Without good conversations, basic mainstays of our health are deregulated, from the immune system to a whole range of metabolic processes that include even the expression of our genes. The cognitive control system with which we govern our ideas and emotions also degenerates. Loneliness turns out to be, therefore, one of the most harmful and ignored risk factors to physical and mental health.

Just as the absence of conversation can have harmful effects, so its presence can be an antidote to certain poisons in our emotional lives. From intuitions forged over centuries, in philosophical thought and in fiction, to cutting-edge science, it has been demonstrated that good conversation improves our ability to regulate emotions. The simplest way is to use words to distract us when an emotion clouds our judgment. Another option is to use our own or someone else's words to induce an emotional state. The most powerful and least known tool is to coin more precise and appropriate terms to describe and reappraise emotions in a freer and less pixelated landscape. These distinct uses of the power of words divide emotional regulation into three broad categories: distraction, induction, resignification. We will enter into the intimate recesses of the brain to understand when and why each method is most effective.

We will discover that the regulation of emotions depends on the same control system that governs attention and thought, which is a clue as to why it is so ineffective to try voluntarily to repress an emotion. It is like wanting to silence an idea: the mere fact of pronouncing it internally, even as part of an attempt to silence it, activates it. This reveals why distraction can be effective in the short term, but it leaves traces that can be measured in bodily indicators of stress. On the other hand, reappraising an emotion instead of suffocating it allows us to link it to other brain circuits and modify the way we experience it. This is the most effective way to use the power of words to govern our emotional lives.

Michel de Montaigne was the last person to speak Latin as his mother tongue. His father created a world for him in the confines of a castle where everyone spoke it, as if that were the most natural thing in the world. That was his own particular *Truman Show*. Beyond the castle walls was a dark world of pestilence and slaughter. Raised with this vast culture yet disconnected from reality, Montaigne lived in complete solitude until he met Étienne de la Boétie. Finally he had someone to talk to as an equal, a soulmate. However, his good fortune was short-lived. De la Boétie died young and Montaigne, who was a prodigy of conversation without an interlocutor, locked himself up in the castle for eight years and talked to himself. That marked the birth of the literary essay.

The right word, the perfect smile

Loneliness has nothing to do with the number of people around us. We can feel isolated in a crowd or when we have thousands of followers on social media. Loneliness is having no one to talk to. This, in turn, as Iranian neuroscientist Bahador Bahrami discovered, atrophies the brain regions that regulate social cognition. This is another reflective principle within the brain itself: perception of loneliness affects the ability of key areas of the brain to establish social ties. Loneliness begets solitude.

This is just the beginning of a harmful loop. Social deprivation degrades a series of processes that condition our physical and mental health, including failures of our immune response, increased blood pressure, and even alterations in the genes that express our cells. The mental health deterioration is even more striking. Loneliness increases the incidence of depression, anxiety, dementia, and attention deficit disorder. Yet despite all these effects, it remains one of the most overlooked risk factors. In general, we tend to underestimate the value of hugs and good conversations.

One of the most emblematic studies on the effects of isolation was done at the beginning of the AIDS pandemic. At that time the pathophysiology of the disease was barely understood and the enormous variability it presented between individual cases was baffling: some people died almost immediately and others were able to live with the disease for quite a while. Some of the factors marking such dissimilar outcomes were predictable—such as the pre-existence of other immune diseases, heart disease, or diabetes—but an unexpected one also emerged: whether the patient spoke freely about his fears, doubts, and circumstances. This example is particularly relevant because back then—and still today—AIDS sufferers were highly stigmatized. A very early study was conducted with gay men, many of whom not only could not talk about the disease but, just because of their sexual orientation, could not even speak freely about their lives. In the case of AIDS, the loneliness factor affected survival by more than one year on average. Those who could talk about what was happening to them lived longer and better than those who had to keep silent. *Stigma produces loneliness, undermining a sick person precisely at the time when they need the most care.* This link is not mere correlation: the quality and length of life of many patients can be improved simply by opening up an adequate caring channel of dialogue. This simple flow of words leads to the creation of a functional cognitive apparatus that is, in turn, the foundation of a good mental and emotional life.

When someone dear to us gets sick, we tend to seek out distractions. That reaction is partly due to the desire for denial and partly also out of the fear and incompetence we feel around relating to someone at such a dark time in their life. We should remain alert and not let ourselves get carried away by this reflex; we must remember how important it is to be present when a loved one gets sick. There may be no more relevant time to express our love and friendship.

Words heal

Loneliness not only conditions the evolution of diseases, but also their appearance in the first place. The flu vaccine is much less effective for people who live alone. Immunity functions within cells, in the expression of genes and proteins. This microscopic mechanism is affected by something that seems very remote: the words a person finds to eradicate possible toxicities in mood, emotions, and desire.

What is the bridge between such different scales, between words and social bonds on the one hand, and the molecules of life on the other? How is it that loneliness triggers a whole series of processes that affect the immune system, vascular function, and even the volume and shape of the brain? While the precise order of this domino effect is difficult to explain, a few studies have revealed the main reasons why social deprivation produces this wide repertoire of physiological problems.

The first reason is very simple: talking to others allows us to make better decisions in the treatment of disease. It is one more example of what we saw earlier: our health decisions are plagued by errors and biases. We tend to ignore and confuse symptoms, be slow to act, fear intervention, incorrectly evaluate who is the most competent person to solve a problem, and misinterpret something the doctor has told us.[30] Each of these mistakes results from the rash decisions we make when we cannot clearly lay out the arguments on the table.

Conversations with other people help us better solve logic problems and make good decisions. For the same reasons it also allows us to make better medical decisions. This is the simplest—not to be confused with the least important—aspect of the matter.

Loneliness also degrades the brain's *control* or *self-regulation system*, a network distributed along the frontal and dorsal lobes of the brain that controls our ideas and goals. The best demonstration of this deterioration comes from an attention experiment known as the "cocktail party effect." We can all relate to the situation of talking to someone in a crowded party, in a room filled with other simultaneous conversations. The problem starts when someone mentions our name or laughs really loudly or talks about hot topics. Those things are strong attention magnets, and set off a mental battle that clearly exposes the mechanism of control and regulation: one part of the brain concentrates on the pyrotechnics of the neighboring conversation, and another tries to ignore them and pay attention to the person we have in front of us.

John Cacioppo, a professor of social psychology at the University of Chicago, designed a lab version of the "cocktail party" in which each ear listens to a different voice. It is not a fair fight: almost all of us are more sensitive to the information we receive through one ear: people who are right-handed favor their right ear and lefties favor their left. Changing this predisposition is complicated because it is an automatic feature of the attention system, just as we find it

30 Once, a traumatologist recommended I rest. When I asked him for how long, he looked at me with an air of confidence and said: "A short while." Of course, his concept of a "short while" and mine could vary by a few days or even many months. Another common version of this variance is with the informal use of probabilities. One person's "very likely" or "unlikely" can mean completely different things to another person, giving rise to contention and calamitous decisions.

very difficult not to direct our eyes over to someone who has started laughing, or someone naked,[31] or the lane with a three-car pile-up. Of course, there are those people who are particularly good at controlling their attention, both in the visual and auditory world and when it comes to shifting focus away from pain, fear, or an obsessive memory. All of these elements are expressions of the control system that allows us to regulate and manage our mental experience. Attention is one of its central roots. Cacioppo demonstrated that the ability to direct our attention at will deteriorates when we lead a lonely life. Loneliness causes diseases, above all because it destroys our system of regulation and cognitive control.

31 When the party really heats up.

Hulk, or the caricature of wrath

The best way to understand the brilliance of emotional regulation is when it disappears. But we can't force it away, so exploring that lack of regulation requires a simulation. This is how fiction has emerged as an extraordinary laboratory for studying the human condition.

When comic-book master Stan Lee came up with the character of Robert Bruce Banner, inspired by Stevenson's novel *The Strange Case of Dr. Jekyll and Mr. Hyde*, he offered us a caricature of a conflicting and painful trait. What would life be like if you couldn't control your anger?

The gamma rays that passed through Banner gave him an amazing quirk. He is a kind and gentle person until his emotional stress crosses a certain line. Then he loses all control and transforms into a beast. The Hulk is the poster child for lack of self-regulation, someone so possessed by anger that he even changes color. Anger not only upsets him, it transforms him, as indicated by his famous phrase: "You wouldn't like me when I'm angry." After the point of no return, he is unable to control his emotions and ends up hurting people he doesn't want to hurt, including himself.

The Hulk is a caricature that reveals anger's most distinctive traits. First, the acquisition of extraordinary bodily powers in preparation for battle. This is wrath's "positive" side. Thomas Aquinas noted that anger arises to eradicate fear and turn it into audacity. On anger's "negative" side we see—in addition, of course, to the persistent damage it causes—ugliness. The Hulk's snot-green wrath illustrates the flip side of a motive that led the ancient Greeks to live a virtuous life: the regulated life is beautiful. The expression *kalos kagathos*, which loosely means "the good and the beautiful," mixes integrity and beauty as two almost inseparable concepts. This association was not invented by the Greeks and isn't unique to them. In fact, we still conflate these concepts today. Control is always admirable: a juggler who keeps twenty pins in the air, the cyclist who rides on the edge of the cliff, or those people who keep their cool in dire circumstances. We find control astounding and beautiful.

The exploration of emotional regulation's limits and consequences is a basic fuel of fiction. The literary critic Parul Sehgal exemplifies this in the following way: "The novel is the lab that has studied jealousy in every possible configuration. In fact, I don't know if it's an exaggeration to say that if we didn't have jealousy, would we even have literature? No faithless Helen, no *Odyssey*. No jealous king, no *Arabian Nights*. No Shakespeare. [. . .] No jealousy, no Proust." Sehgal argues that jealousy is not only the omnipresent subject of literature, but it also makes us novelists: from a mysterious phone call or someone coming home later than expected, the jealous create a plot full of convoluted details in which they are the only victims. And we are particularly good at finding arguments that we can bring together in impeccably coherent narratives, even if the story bears no relationship to reality. And, of course, we end up believing those extraordinary stories we tell ourselves. Understanding that jealousy inspires fiction is a first step for those who want to govern their green-eyed monster. Just as writers choose what they want to write about, we have the freedom to script the story we tell ourselves. And, depending on how we frame that story, we are able to transform a feeling of jealousy into one of tranquility or confidence.

In the wide range of literature that deals with emotional regulation, the Hulk is just one contemporary example. Back among the first stories and poems, we find Ulysses. As he sets out on his journey back to Ithaca, Circe warns him that he will sail by the sirens, whose song is so magnetic that it drives any men who listen to them crazy. Desire sets such traps for us. The sailors plug their ears with wax and Ulysses orders them to tie him to the mast.[32] The sailors' solution is more straightforward, yet effective: ignoring temptation. Ulysses, however, does not want to silence the singing. He knows and accepts his own limits, the contours of what he can and cannot do. The "Ulysses pact" as this strategy is still known today, is a commitment to the future decided in the serenity of the present. An effective way to resolve the traps of the *Odyssey* that are actually mind traps.[33]

---

32 Ulysses should have given his crew their orders before they had wax in their ears.

33 In 2004, Alessandro Baricco dismantled and reassembled the *Iliad* for a public reading project. He kept the scenes and images, but eliminated the voices

Most of us find it difficult to stop glancing at our phones frequently, as if a notification that will change the course of our lives could appear on the screen at any moment. Of course, that doesn't usually happen, but the illusion persists and so we often allow important moments of our lives to be interrupted by this futile temptation. We are having dinner with a friend who sits there right in front of us and we discuss intimate topics, but hovering over that private space is the threat of the two mobile phones sitting on the table, which, like the sirens of Ulysses, emit irresistible songs. The same thing can happen during the two minutes when our child is confiding in us about his fears, or telling us about the best part of his day. I've been leaving my phone in another room during family dinner for years. It is my Ulysses pact, made in the serenity of the present, recognizing my own inability to rein in the impulse to check it when it's in arm's reach.

Emotional regulation

These ancient ideas remain vital in the most modern versions of emotional regulation, an incipient science built from old traditions,

---

of the gods. "The *Iliad* has a strong secular skeleton [. . .] Behind the gesture of God, [. . .] it almost always mentions a human gesture."

from Homeric narratives to psychotherapy, which merge into cognitive science, experimental psychology, and neuroscience. This project has revealed the following tools for improving our emotional lives:

1. **Distraction**: use words to distract ourselves when an emotion overpowers us and takes up all our mental space.
2. **Induction:** use our own or others' words to induce emotional states. Use a story to elicit anxiety, joy, or fear.
3. **Resignification:** coin more precise and appropriate words to be able to describe emotions in a less pixelated landscape.
4. **Compassion:** converse to make better decisions, and avoid loneliness and the illness it can provoke.

We will break down each of these strategies to explore where and why they are most effective and powerful. But first let's see them all together, condensed into a single scene: the day you become a hero.

The prelude to the abyss

We are in the semi-finals of the 2014 soccer World Cup. In Brazil, Argentina and the Netherlands finish the match nil–nil and go to a penalty shootout. Javier Mascherano, captain of the Argentine national team, embraces goalkeeper Sergio Romero with these words: "Today you become a hero." Just five words spoken with a precise tone and a searing look at the right time. "Chiquito" Romero stopped the first and third penalties of the shootout, two out of three, something that is far beyond what usually happens, and led the Argentine team to the final. That day, Romero became a hero.

There are many more examples. The defining ball that always reaches Michael Jordan in a championship with millions of spectators, amid many rivals warned that it will be Jordan alone who decides the outcome of that decisive second. Katie Ledecky, poised at the Olympic final in Rio. She looks like a carbon copy of the other female Olympic swimmers, but no. She swims four hundred meters faster than any other woman in history and manages an unattainable distance of several body lengths from her competitors. New York, 1999: on one side of the net is Serena Williams, barely seventeen years old; on the other, Martina Hingis, the number one player in the world. The match is resolved in a few seconds, in a tiebreak. It's Williams's serve with a six–four lead, she keeps her cool and wins the first Grand Slam final of the twenty-three she will take home over nearly two decades spent dominating the world of tennis. Rafa Nadal in the 2008 Wimbledon final. Over the more than four and a half hours of the most epic match in tennis history, Federer had thirteen break points. He only won one. Johannesburg, minute 116 of the 2010 World Cup final. Three minutes before going into extra time, Andrés Iniesta receives a pass in the penalty area. When recalling the most important goal in the history of Spanish football, Iniesta says: "In that instant, as soon as I received the ball, there was

a sudden silence. There were thousands of people in the stadium, but *in that moment it was just the ball and me. The two of us alone.*" Freddie Mercury steps onto stage in front of a vast sea of people waiting for him at Wembley. He's not intimidated by the crowd. It energizes him and stimulates him to give the concert of his life.

Jordan, Ledecky, Nadal, Iniesta, Federer, Mercury, and Williams are revered for their talent, effort, and skill, but perhaps above all for that unique mettle that allows them to perform in the most difficult moments. They are fabulous at controlling their emotions. They manage to channel them when the rest of us fail. Just as we can learn to regulate emotions by studying their exaggerated failures—in the case of the Hulk—we can also learn from the natural wizards of regulation. How does an Olympian manage to fall asleep on the night before the competition? How do you reach maximum concentration at the decisive moment? How can you manage to persist, battle after battle, without wavering? How do you maintain your ambition after reaching a great milestone?

The goal of this analysis is not for us to become Olympic athletes by managing our fears and anxieties. It is more modest and realistic; we can become heroes without famous exploits, without television fanfare, simply in the future of each of our own lives. It is enough that we finally manage to say what we couldn't before, that we resume a project we were so excited about, that we overcome bashfulness and tell that person how much we love him or her.

Prepare, dodge, reflect, attenuate

I have talked to many athletes in search of answers to these questions. Four paths systematically appear, each of which corresponds to the pathways of emotional regulation laid out above.

First, there are mantras, stirring speeches, and rituals. Those include, for example, the famous *haka*—the Maori ritual with songs, yelling, and warrior stances that the New Zealand rugby team performs before each match to embolden themselves and intimidate their opponents—and those five words spoken by Mascherano. Rafa Nadal, one of the greatest tennis players in history, performs a series of ritual tics before starting each point: he sweeps the line, adjusts his shirt and then his shorts, wipes his nose, arranges the hair behind his left ear, wipes his nose again, and then tucks his hair behind his right ear. That sequence puts him in a state of mind conducive to maximum concentration and performance.

It seems surprising that an emotional state can be induced, sometimes so easily. However, it is very common and simple; just evoking a cheerful image can make us feel good, even if only for a second. Emotional regulation begins, as in the Ulysses pact, before we experience an emotion. We will see that this tool has its limits and handicaps.

A second way to deal with emotional upheavals is distraction. A well-paced film or an addictive series is capable of shifting our mental narrative away from the monolithic discourse we construct when something is worrying us. Distraction on social networks is even more common, although it incurs the risk that we will find precisely what we were looking to escape from.

Other times, emotional control does not stem from the prior induction of a resilient state of mind, nor from a distraction by stimuli that compete with our fear. This third strategy is quite different because it doesn't try to turn off the rushing sensations, but to transform them into an invigorating stimulus. Marvel has its caricature of this in the mutant Sebastian Shaw, an adversary of the X-Men, who manages to absorb all the energy directed against him and transform it to multiply his strength. In some martial arts, such as judo, the attack is also nourished by the opponent's strength and weight.

This is perhaps the most interesting and powerful form of emotional regulation. This is what is known as *resignification.* It is about shifting the course of, without interrupting, our butterflies in the stomach, trembling, or increased heart rate. After all, the body reacts automatically and explosively because something very important is about to happen. Sebastian Shaw's power is to turn all these sensations into something that gives us strength. Many times it is enough to change the word, give another name to that set of sensations. Instead of automatically saying we are afraid, we could instead say we're excited.

A single sensation can be interpreted with nearly opposite connotations. Two different terms—"fear" and "thrill"—are often used to designate the same emotion. In fact, Google defines "thrill" as "a sudden feeling of excitement and pleasure" and gives the example "the thrill of jumping out of an aeroplane." I confess that in my case before jumping out of a plane I would be feeling absolute panic, not excitement and pleasure.[34] This ambiguity gives us the freedom to become Sebastian Shaw, to reappraise fear as the declaration of something imminent that is fabulous, unique, exciting. This group of bodily sensations are the most intense proof that we are alive. Like feeling the wind in our faces.

Let's have a look at the fourth and final category of emotional regulation. A tennis player makes a huge mistake at a key point, very close to the end of the match, and begins to disparage himself with phrases like: "All my life spent playing tennis and I never improve!" Their coach and the crowd do something very different. They seek

34 The etymology of panic refers to the god Pan, known for his noisy antics, and to whom unexplainable rumbles of the earth are attributed. Panic in the air?

to encourage the player with phrases such as: "That's OK, you can do it! Hang in there!"

We usually speak to others with a compassionate, embracing voice, yet the voice we use to speak to ourselves is often judgmental and accusatory, heaping insult onto injury. The sphere of "self" extends to our closest circle. In many cases, the most harmful place where this vice manifests itself is with our children. There is nothing I want to change more than my urge to get angry with them when they stumble, my impulse to lecture and upbraid them for being so distracted or careless. I am wholly convinced that this is the time to embrace rather than judge. Despite that, I sometimes forget. For me, the most beautiful thing about a regulated life is banishing the Hulk response when what the person I love really needs is a hug.

There are no magic recipes for schooling ourselves in these principles. Just as there is no phrase that can *ipso facto* turn us into great poets or tennis players, there isn't one capable of suddenly transforming us into great navigators of our own minds. But we can practice and improve, and many times even a small change is enough to give our lives a more interesting and enjoyable tint. That is the raison d'être of this book, which began with a personal journey, with my desire to moderate some of the excessive fear, anger, lack of compassion, and jealousy that sometimes disrupt the most beautiful aspects of my life. I suspect we all are similar in that way.

Implant an emotion

To understand induction, it is useful to review the nature of emotions. An emotion is a precise conglomeration of experiences: our conscious sensation of it, our physiological responses to it, the gestures we use to communicate it, and the events that trigger it. We already saw this with *liget*, whose peculiarity was not in each of the separate features that defined it, but in the whole.

Once the mixture of expressions that accompanies an emotion is discerned, the next question is the one the great American psychologist William James asked some one hundred and fifty years ago: Which comes first, the sensation or the physiological changes? It would seem that the emotion—anger, joy, or sadness—is experienced first and then expressed, in the same way that we can only communicate an idea after formulating it. However, that is not the case. The brain reads bodily states to discover or build the emotions we experience. We laugh, *ergo* we are happy; if we grit our teeth, we must be angry. Just as we recognize others' emotions through their body language, our brains also use our bodies to infer our own emotions.

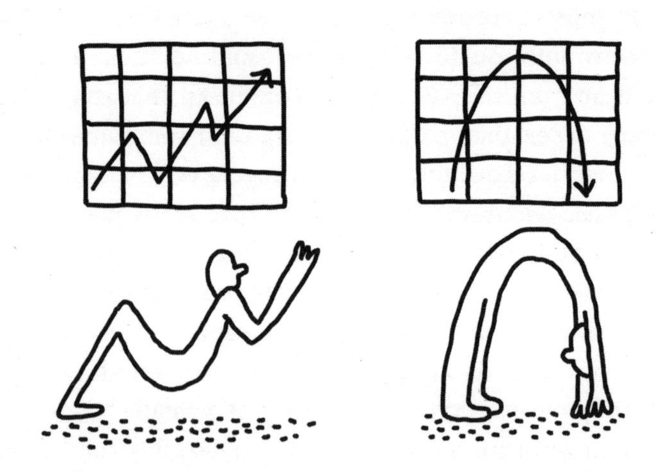

The expressions of an emotion form a feedback loop that is, by its very nature, reflective. The spark can be ignited at any point. It can start with the feeling evoked by some news we receive, which in turn triggers inconsolable crying. But it can also happen the other way around. If someone laughs in the middle of a crowd, the brain can register it (without our being conscious) and—through the mirror neuron system, which automatically imitates the reactions of others—set off the motor signal to produce a smile. This in turn generates a feeling of happiness not elicited from any news, but from the induction of bodily expressions of those around us, by mere contagion.

The prime example of this idea comes from the famous experiment carried out in 1988 by German psychologists Fritz Strack and Sabine Stepper. They had the participants watch a series of cartoons and evaluate how funny they found them. Some of the subjects held a pencil between their lips as they watched, and others held one in their teeth. This subtle difference completely changed the configuration of their facial muscles: holding a pencil in our teeth makes us look as though we are smiling, whereas holding it between the lips produces a look suggestive of anger. Strack and Stepper's results show that images are considered funnier when judged by those holding the pencil with a smile-like expression. This is a simple and powerful demonstration of how the mental experience can be tinted. The world seems more fun when seen through a muscle *filter* that widens the mouth to endow the experience with a faint smile.

There is an even simpler and more relevant version of the Strack and Stepper experiment that anyone can carry out. Just hold a pencil between your teeth, and you will almost immediately perceive a feeling of joy. This is the inductive effect with which gestures are able to implant an emotion. Of course, there are limits to this. The joy fades quickly and keeping it alive requires a muscular effort beyond what we feel when we laugh naturally and spontaneously. The induced emotion, moreover, doesn't have the precise tone of joy and feels somewhat imposed. Thousands of continuations of Strack's experiment have been carried out in universities. One of them resonates with what we've already seen about the effects loneliness has on health. Psychologists Tara Kraft and Sarah Pressman showed that mimicking a laugh not only produces a sense of joy (as anyone can see) or makes us find things funnier (as Strack showed): it also improves our physiological response to stress. In other words, to some extent, it is curative.

One of the most important continuations of Strack's experiment was made thirty years later. During a revolution in social psychology known as *the replication crisis*, scientists set out to review a cluster of experiments that had become dogma and yet were based on scarce and sometimes also biased data. Strack's experiment was repeated in seventeen countries with a sample twenty times larger than the original, and this time the entire process was videotaped. With these rigorous controls, the induction effect disappeared: the cartoons were just as funny when the pencil was held in the subject's teeth, forming a smile, as when held between the lips, producing an angry expression.

Even though they were still cited, Strack's original conclusions seemed to have been definitively refuted. A couple of years later, however, Tom Noah and Ruth Mayo discovered that the discrepancy between the two versions of the experiment was due to an unexpected effect. The *fault* lay with the camera that was used to better control the process. Facial gestures influence how a story is perceived, but the reflective power of a smile is fragile and the mere presence of a camera ruins it. There is an everyday example where this becomes apparent: when we are asked to smile for a photo. Rather than joyful, what we feel is overwhelmed and uncomfortable. It's the same facial expression, the same gesture and yet it yields a completely different result. The context is a determinant and that is, precisely, the second formula in the process of inducing an emotion.[35]

We've looked at three experiments, each of which has a different scope. First is the one that any of us can do by biting a pencil to induce an emotion; then there is Strack's, which showed that a simple muscular expression allows us to change the way we perceive the world; and finally Kraft and Pressman's, which shows that consistently turning that frown upside down is a good antidote to stress. This links induction to distraction, and so we begin to see how the premises of emotional regulation can be combined.

Along the way we also see an intrinsic limit to this form of emotional regulation: the happiness induced by a fake smile is ephemeral. If we want to think of induction as a tool to regulate emotional life, we must address this issue. The simplest possibility is to use it only when we don't need the effect to last long, the way Nadal uses his ritual tics to reach a maximum state of concentration just before his serve. A second, more challenging, possibility is to think about how we can induce more enduring emotional states. The solution is found in a version of Strack's experiment, which is well known to every stand-up comedian. While it is hard to ignite

35 I included the details of this story because it exemplifies the complexity of the scientific process. Premises and experiments are, often, simpler than the interpretation of their results. Especially in a realm as rich and variable as the human one. Science does not preach truths, it only offers approximations of reality, which we would be wise to examine with healthy skepticism.

the spark of laughter, once it is lit, the fire spreads. The same joke lands very differently depending on whether it's told before or after that turning point: in one case, roars of laughter; in the other, nothing but silence. Why does our perception change so much? The answer is contagion: the loop is no longer between one person's brain and body, but an entire group's. And this adds a second resonance, like a choir singing in an echoey church. Induction, which predisposes a person to chuckle, has the same effect on the person next to him, and so on and so on. We are the kindling that makes laughter spread like wildfire.[36]

36 In 1994, Hortensia Gutiérrez del Álamo's laughter spread throughout the Andalusian parliament so infectiously that its president, Diego Valderas, was forced to suspend the session. It is impossible to watch that video without contagion.

Strack's experiment allows us to come up with a simple but effective formula for everyday life: try to be surrounded by good-humored people. Increased spontaneous laughter produces general well-being and improves many health indicators, the same ones that anger puts at risk. Erika Rosenberg showed that the problems of heart disease patients are aggravated if they live with bitterness or are afflicted with resting stink face. The bodily expressions of the people with whom we live mimetically induce our own expression and this in turn—as at a comedy show—changes our perception of the world.

Distraction and attention

Distraction is the most intuitive tool for emotional regulation. It is also the one requiring the least effort and, as such, the one that is both most used and most abused. Distraction was among Aristotle's recommendations, although I won't delve into its history here. Rather I'll discuss the most recent discoveries regarding distraction in order to address how to employ it most effectively.

Often we find ourselves distracted without any effort on our part. While reading, for example, it is common to suddenly realize that we've disconnected from the book.[37] Our eyes sweep over each word, even slowing down their movement at complicated passages, while our mind wanders at will. When you reach the last word on the page, your arm extends, your fingertips gently flip the page, and your eyes again sweep over, word for word, the new paragraphs on the next one. Your mind, however, is elsewhere, to the point that you have no idea what it is you just read. This is a fabulous example of distraction. In full wakefulness, the mechanics of the body are separated from conscious experience. Our eyes register the words while our consciousness is inundated by daydreams. We don't usually think of it that way, but it is a fabulous power that allows us to completely disconnect from our sensory experience. The problem with that power is its unpredictability. It is almost impossible to control these daydreams. If we were to *try* to distract ourselves from reading, we would never be able to detach ourselves enough to stop processing the words.

---

37 Can you summarize what you read on the previous page?

Michael Posner, one of the pioneers of cognitive science, exhaustively broke down the mechanisms of *attention*, another concept whose granularity often confuses us. We use it colloquially to refer to the need to concentrate in class, to be alert or vigilant, to highlight something that has been said, or even when someone takes no interest in us, saying that they "aren't paying attention to us." Actually, each of those cases refers to a different kind of attention, each composed of many pieces.

Posner applied science to the study of attention, meticulously, and managed to identify its four main mechanisms:

1. **Exogenous orientation.** A door opens suddenly, we hear a gunshot, someone touches us or calls out unexpectedly, something falls out of our pocket, a pedestrian walks out in front of our car. In each of these examples, our mental focus shifts automatically. This is an exogenous way of directing attention, which is one of the functions of the attentional system.
2. **Endogenous orientation.** This is the ability to direct your attention at will. We want to read a sign in the distance and we focus all of our visual effort on it. A driver dealing with city traffic for the first time, summoning maximum concentration. We can compare this state of mind with the one the same driver would have, years later, taking the same route, which they now know like the back of their hand, while immersed in daydreams. We employ the endogenous orientation of our attention when doing anything for the first time, which is why first times are so memorable.
3. **Sustaining attention.** A student in a math class is focused but eventually starts to get bored. That is the beginning of a struggle to keep their attention from straying, led by this basic mechanism of the attentional system.
4. **Disengaging attention.** An obsessive idea, an argument that just adds fuel to the fire, a whim, a game, everything

> we recognize as addictive. Sometimes the mind gets trapped in a loop and the attentional system has a mechanism to rescue it from continued spiraling.

Posner discovered that each of these functions involves independent brain systems that develop at different times in life. The first to activate is the system that allows attention to be oriented exogenously, while the networks that regulate our ability to detach our attention take much longer to mature. A clear example of this lag is first-time parents responding to their baby's sustained crying by begging him to stop, until they discover, through trial and error and sleep deprivation, a much more effective trick: offering the baby another stimulus to attract his or her attention. And then, as if by magic, the crying stops. On many occasions (although not always, of course) the persistent crying is explained by simple inertia. At that age we are as capable of focusing our attention on an exogenous stimulus as we are incapable of voluntarily detaching it.

Learning the components of thought development helps us to have more satisfying relationships. No parent would ask their six-month-old child to run. For the same reason, knowing how our attention evolves as we grow can also prevent a parent from asking their baby for the impossible: to just stop crying on command, at will.

As with almost all cognitive faculties, those that develop earlier are more persistent and leave traces that are expressed throughout our lives. The asymmetry within the attentional system improves with age, but still persists. It is not easy to draw our attention away from something that obsesses us, hurts us, or irritates us deeply. Distracting ourselves is still much easier.

This brings us back to the idea of our minds wandering as we read. Distraction is rudimentary when we muster it voluntarily. On the other hand, it becomes much more effective when drawn to a magnetizing stimulus. Which stimuli have this powerful magnetism? Sugar, drugs, pornography, television, social networks, video games.[38] There is no short supply. Sometimes, to distract us

38 On Twitter this analogy went viral: Tinder—Lust, Instagram—Envy, Amazon—Greed, Twitter—Wrath, Netflix—Sloth . . . Each app has its cardinal sin. This logic is a guide for those who only invest in a company if they detect that their products are fuel for one of those vices.

from pain, sadness, and fear, we need such efficient fuels to divert our attention that the remedy ends up being worse than the disease. This is why, when we are suffering from depression, anxiety, or stress disorders, our health can decline even further with these addictions.

Anyone who has experienced very acute pain knows that there is no stimulus, however powerful, that can distract from it. That is when we realize that this form of mind control has its limitations. That small gesture of humility can save us by recognizing when the time has come to try other tools to regulate our feelings. The next level is based in words and can be incredibly powerful. It is about resignification: the ability to change our interpretations of our feelings, to make them more palatable.

The construction of fear

When my nephews were eight and ten years old, we traveled together from Buenos Aires to Madrid. At some point, past the north of Brazil and already over the Atlantic, we went through a turbulence that began to shake the plane violently. I grabbed the armrests[39] and wished with all my strength that it would end soon. Then I remembered my nephews and thought they must be terrified. I took a deep breath so I would seem as calm as possible and turned to them to convey my feigned serenity. That was when I saw that they were waving their arms, jubilant, and shouting, "Roller coaster, roller coaster!"

During a roller coaster's free fall, our viscera move to our lower belly, our heart rate shoots up and we start screaming in a panic. Why do we come back for more? Why do we pay to experience fear? Precisely because the roller coaster is where we discover that fear can become pleasure.

Since that trip to Madrid, the image of my nephews waving their

---

39 I had won them in what Christoph Niemann calls the fight for "armrest supremacy": "You know, if I want to tell the story of modern-day struggle, I would start with the armrest between two aeroplane seats and two sets of elbows fighting. What I love there is this universal law that, you know, you have 30 seconds to fight it out and once it's yours, you get to keep it for the rest of the flight."

arms at ten thousand meters in the air has become a kind of mantra for me: "Roller coaster, roller coaster." It's no magic potion, and it doesn't work immediately, but it does work. Every time the plane falters, I imagine I'm a passenger in the early days of aviation who has waited in line to get on that fabulous attraction, a mastodon with wings soaring through the sky, and I feel much better. Sometimes I even manage to enjoy it. This formula has also worked for me in many other contexts and situations, especially those that evoke visceral reflexes and fears without actually being very dangerous. For example, when I had to give a very important talk or play in my first concert.

How is it possible that we can turn fear into pleasure? As compared to distraction, resignification is mysterious and much less intuitive. And that is because it requires unlearning deep-rooted associations: our bodily experiences of fear, anger, and sadness are so linked to their sensations that they seem inseparable. Yet they aren't. Let's look at why that is, beginning with a unique example.

The famous mountain climber Alex Honnold is often asked how he conquers fear while climbing without ropes along vertical walls and hanging from tiny outcrops hundreds of meters from the ground. But of course, before attempting those things, Honnold made many of the same climbs with ropes, so he could estimate their degree of risk with great precision. The unique thing about this climber is not the enormous challenges he faces, but the extraordinary technical virtuosity that allows him to reduce the risk of these ascents to a reasonable range. Actors lose their fear of the stage, Formula One drivers lose their fear of speed, pilots lose their fear of turbulence, and a surgeon becomes inured to the impact of seeing bleeding bodies. In the same way, an expert climber makes walking on cliffs into a habit that, with practice and experience, reduces their measure of risk.

Fear of heights seems to be universal and innate, rooted in our genes as a basic principle of survival. But it's not. Like so many other emotions, we learn to imbue it with meaning. This was discovered by the psychologist Karen Adolph, through curious experiments in which she made very young children walk over bridges placed at different heights.[40] Adolph discovered that fear of heights is learned with experience. Babies first fall and only then learn to fear falling, not the other way around.

When babies start to crawl, they approach heights without a hint of fear. Just a few weeks later, when they've acquired experience of moving through the world on all fours, they begin to show signs of understanding the risk of the abyss. They'll stop at the edge and explore it. Over time they become more cautious until, almost with adult precision, they can sense whether or not they're prepared to overcome an obstacle. It seems they've learned the fear of heights, but that's not actually the case.

Months later, when they start taking their first steps, the process repeats itself. In those first few times they will, once again, walk over any cliff, as if everything they've learned crawling is forgotten as soon as they begin to walk. And again, as the weeks go by, they begin to walk much more carefully and attentively. When they start walking, babies fall about forty times an hour. They almost always get up quickly and keep playing, as if nothing has happened. Falling is the natural mechanism for discovering and learning about danger. We estimate the probability of falling with each movement and, based on this measure of risk, we construct our fear.

The vertigo that the vast majority of adults feel at great heights is acquired over time. Honnold is unique, but not in his ability to keep his cool at great elevations. There are many people who can do that: trapeze artists, pole vaulters, paratroopers . . . All of them have learned something that seems impossible to the rest of us: to consider heights with their appropriate measure of risk, the way most of us do, for example, when crossing the street.

Our minds can also be tricked in the opposite way: when we don't respect the true risks inherent in dangerous situations. In this case, I will use an example from my own experience. In my twenties I traveled to Tayrona National Park in northern Colombia,

40 At least half of the children survived these experiments. Just kidding, of course. Actually, the survival rate was over 90 percent. Just kidding. There was no risk to the participants. The only thing at risk is our sense of humor.

one of the most beautiful places I've ever seen. In Tayrona you can walk for entire days through jungles, along tropical beaches and amid archaeological relics. Throughout much of that dreamlike landscape, coconuts rain down from several meters high. No one there seemed bothered by the risk of those heavy, falling spheres. Except for me. To me it felt like walking down a street where people threw stones from their balconies. It was definitely a similar risk, and yet everyone there took that risk for granted; the magic of the jungle hid that danger and imbued it with new meaning. So effectively that, when I approached the park ranger to ask him about the coconuts, he brushed me off with something that was clearly a common phrase: "The coconut knows when to fall." I have a vivid memory of the contrast between his composure and my agitation. "The coconut knows!" I went back to Santa Marta, at the entrance to the park, where I got my hands on a hard hat, the kind construction workers use. I wore it throughout my explorations of the Tayrona jungle, looking ridiculous but feeling happy. When I got back home, curiosity (and perhaps hoping to find arguments that would make me less alone and ridiculous in my fear) led me to investigate the subject, and I found that the coconut doesn't always know when to fall and actually causes a good number of accidents, many of them deadly, as P. Barss points out in a famous article published in the *Journal of Trauma* titled "Injuries Due to Falling Coconuts."

These cases offer opposite examples of the same phenomenon: how we construct stories to modulate our fear. We are capable of heightening fear even when true risk is negligible and, on the contrary, able to dispel it despite high-risk situations. That is why there are so many rare phobias that most people find incomprehensible.[41] A fabulous exercise in empathy and understanding consists

41 In *Cronopios and Famas*, Julio Cortázar explains it like this, in Paul Blackburn's translation: "At different times the family has tried to get my aunt to explain with some sort of coherence her fear of falling on her back. On one occasion the attempt was received with a silence you could have cut with a scythe; but one night, after her glass of sweet wine, aunt condescended to imply that if she were to fall on her back she wouldn't be able to get up again. At

of not judging or dismissing others' fears and, taking it even a step further, in understanding that the terror is always real to those who are experiencing it. When we keep that in mind, caring for and protecting others becomes much more natural.

which elemental observation, thirty-two members of the family swore they would come to her aid; she responded with a weary glance and two words: "Be useless." Days later, my eldest brother called me into the kitchen one night and showed me a cockroach which had fallen on its back under the sink. Without saying a word, we stood and watched its long and useless struggle to right itself, while other cockroaches, prevailing over the intimidation of the light, traveled across the floor and passed by brushing against the one who was lying there on its back waving its legs. We went to bed in a distinctly melancholy mood that night, and for one reason or another no one resumed the questioning; we limited ourselves now to alleviating her fear as much as possible, escorting her everywhere, offering her our arms and buying her dozens of pairs of shoes with gripper soles, and other stabilizing devices. That way life went on, and it was not worse than any other life."

Granularity and ambiguity

We see the same process with the rest of the emotions as we've seen with fear. At the start of this book we mentioned the reason behind this: the reflective power of words. We can turn frustration into anger, anger into sadness, sadness into joy. In each of these examples, the visceral sensations may be identical, as in this ambiguous illustration where we can see completely different things.

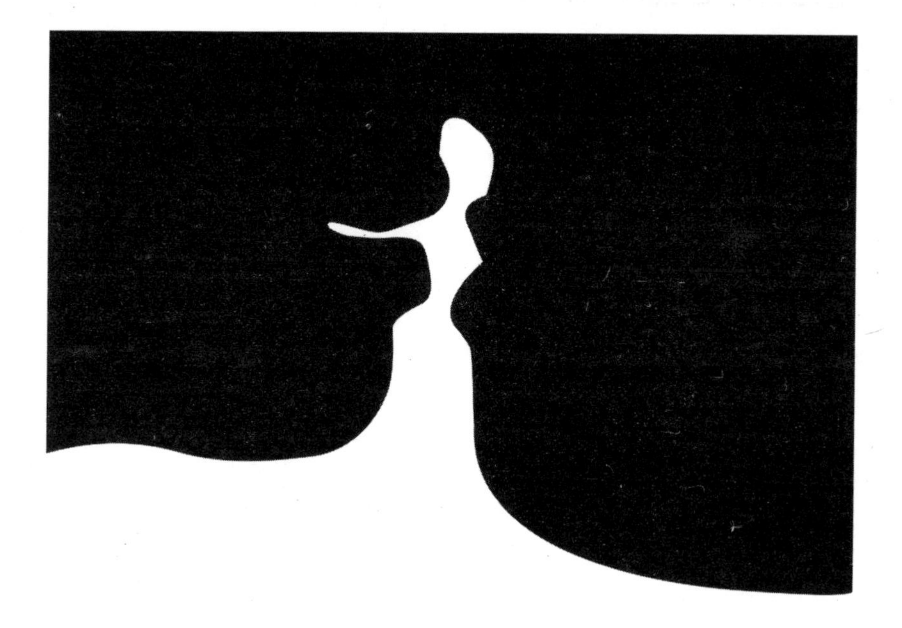

This analogy allows us to identify three principles of ambiguous images, which correspond to the world of emotions: 1) some are much more ambiguous than others; 2) alternating between one interpretation and another requires some learning; 3) each person has their own bias and immediately *sees* one of the two images. The same thing happens in the world of emotions: 1) some people are more likely to get confused than others, as we saw when we explored Plutchik's wheel; 2) reappraising requires significant practice and has a learning curve; we can't just will it to happen; 3) each person has their own predispositions and there is usually a feeling that dominates their emotional experience.

The last principle explains the arbitrariness of how we each regulate our emotions. Should we turn sadness into anger or turn anger into sadness? There is, of course, no single right answer to this question. Rather it seems to be governed by Paracelsus's toxicology maxim: the dose makes the poison. The same can be said of emotions: each person must find their proper dosage.

Surely the Hulk, and those who identify with his excesses, would strive to abate anger. On the other hand, there are some people who instead seek to transform their emotional lives so as to give anger a free rein. Writer and activist Soraya Chemaly embraces anger as a powerful emotion that has been stigmatized in oppressed collectives. Chemaly says that, in the United States, a Black person's anger is interpreted in a radically different way from a white person's, and that girls are taught that anger is masculine and sadness is feminine.

It is not a new story. Plato wrote in *The Republic* that the public expression of sadness is not a matter for men, while giving us a recommendation for emotional regulation that remains valid: the theater. "When even the best of us hear Homer [. . .] we suffer along with the hero and take his sufferings seriously," says the philosopher in Reeve's translation, "and we praise the one who affects us most in this way as a good poet." That is, "real men don't cry" in life, but in the theater they find a space where they can exercise that emotion.

Every culture has its designated terrain of emotional expression. Some are typical of a group—children, men, women—and others are defined by our workspace: judges must remain impassive, flight attendants have to radiate calm, and salespeople happiness. Not to mention the professionals: mourners hired to cry at wakes, and television audiences paid to laugh. Other times, as in Greek theatre, the expression of some emotions is only *acceptable* in certain spaces.

It is worth repeating the question, now in the first person: Do *I* want to turn sadness into anger, or anger into sadness? It's worth gazing at ourselves in the mirror of emotions to identify which ones dazzle us and which are conspicuously absent. The purpose of resignification and using words to regulate and train our emotional lives is not to make humankind monochromatic, devoid of anger and sadness. Quite the contrary: it is about finding a more granular representation of our emotions that allows us to see their versatility and how they overlap, and giving ourselves greater control over when and how to feel fear, anger, happiness, sadness, jealousy, and surprise.

The granularity of emotional space is also not a new idea. New York-based philosopher Mariana Noé, who studies the emotions of ancient Greece, points out that Aristotle was the first granular philosopher. Aristotle listed the human virtues and their corresponding vices according to their degree of excessiveness. Friendliness is a virtue flanked on one side by the vice of cantankerousness and on the other by excessive obsequiousness. Courage lies between cowardice and rashness. In the resulting table there is, however, an empty box: between ambition and lack thereof we find proper ambition, for which, Aristotle already noted, there is no single word. The matter becomes even more interesting when translations come into

play. In Spanish there is no suitable term to designate the virtuous center between pusillanimity and vanity, which J. K. Thomson terms in English as "magnanimity."[42] Nor is it clear which virtue lies between stinginess and prodigality, or what is the intermediate point between boorishness and buffoonery. In general, it is difficult to find the single right word to describe a virtue.

The brain that resignifies

Now we will look at how best to take advantage of the science that has flourished in recent years, which shows that resignification is one of the most effective and versatile tools of emotional regulation. We will also learn its limits and idiosyncrasies, so we can use this tool for sculpting our mental experience in the best possible way.

For more than a century, emotional resignification remained in the realm of psychotherapy. About thirty years ago an approach was developed that made it possible to elucidate its mechanisms scientifically. One of the great pioneers in bringing resignification into the field of neuroscience is James Gross, who, with his students and colleagues, has done countless experiments whose results encompass a tremendous encyclopedic manual. I will try to outline the essential ideas that have emerged from his years of intense study.

The typical experiment works like this. In the lab, people look at emotionally intense images or videos. One group simply observes them; another tries to distract themselves from them, and the third is asked to reappraise the emotions observed.

The first finding was that distraction is effective as a tool to attenuate the subjective experience of a negative emotion or pain, but at great expense to the body. Distracting our attention away from an emotion paradoxically produces a more vigorous bodily response, with an increase in heart rate, stress response, and vasoconstriction. This is the physiological proof of an intuition that is by no means new. It is linked to the idea of repression in several psychodynamic

---

42 Thomson's term is close, but seems to allude more to generosity. Another word sometimes used is *pride*, but in almost all its possible meanings it is much closer to vanity than to a true midpoint. What other word can you think of to capture that emotional middle ground?

theories, including, of course, those of Sigmund Freud. Hiding the dust under the carpet works in the moment, but it leaves scars.

On the other hand, resignification produces an equal—or even greater—effect and attenuates the subjective experience without accumulated stress, and without the emotional trigger continuing to reverberate within the body. This also demonstrates the value of these studies, which show that, although the two tools are similar on the surface of the emotional experience, they differ in another set of expressions *invisible to the naked eye.*

We will now travel back into the brain to see what happens at the precise moment when someone tries to change their experience of an emotion. Our goal goes far beyond merely locating the regions that are activated by a certain behavior. This sort of mapping exercise, in my opinion, offers no real contribution. Our objective is to identify, as Michael Posner did with attention, the *functions* that are orchestrated to carry out this cognitive feat. In other words, brain observation is useful when it allows us to discover the basic operations that constitute a complex cognitive process—such as emotional regulation—where the mere observation of verbal expressions and behaviors is not enough to decipher them.

The brain activity of a person who is reappraising an emotion, compared to that of someone simply observing it, presents the following differences:

1. **Decreased activation in the amygdala and a portion of the medial orbitofrontal cortex**[43]
   These two regions index the intensity of an emotion, especially fear. The amygdala has two very different response phases. The first is automatic, like a reflex. It is so fast that it allows the amygdala to encode the emotion of a facial expression before we even detect whose face it is. We know what that person feels before we know who they are. That component is unaffected by resignification; there

43 All we have are these complicated words to refer to the coordinates on the atlas of the brain.

simply isn't even time to make that happen. In the second, slower wave of activation, the amygdala connects with other regions of the brain and prevents them from functioning normally to carry out their usual tasks. The brain is *hijacked* and overwhelmed by emotion. This is what we observe in a fit of anger, when we lose our ability to reason and often even our capacity for perception. This second wave, in which various regions of the brain connect with the amygdala, also gives rise to the formation of the memory engram and, as we saw earlier, it is here that the neural circuits linked in a memory begin to interweave. Resignification modulates precisely this second phase in which the amygdala fires, connects to the rest of the brain, takes it by storm, and forges indelible memories with thousands of connections. That's why it's so effective at putting out the emotional fire before it leaves scars and we start to become, like the Hulk, greener and greener.

2. **Increased activity in some regions of the cognitive control network, particularly the prefrontal cortex**
   The cognitive control network is made up of several regions, including the anterior cingulate and the prefrontal cortices. The anterior cingulate cortex is a kind of monitoring control tower that sets off an alarm when something isn't working. The prefrontal cortex has the ability to coordinate the flow of information between different brain structures. It's like a traffic cop. Cognitive control has clear limits and sometimes the effort to lead thought produces effects different from those intended. The cartoonist Quino captured it very well in that comic strip where Felipe conscientiously repeats: "I must pay attention, not missing a single detail of what the teacher is explaining. I have to concentrate all my attention to avoid distractions, focus all my senses on being attentive." While he is absorbed in these ruminations, he suddenly hears a voice saying, "Is that understood, children?" and Felipe is

heartbroken when the rest of the class unanimously responds: "Yes, Miss!"

The brain's cognitive control network is also activated during emotional resignification. This implies that there is substantial overlap between the brain systems that regulate emotions and thought. They use the same functions, the same circuits, and the same instructions. In the chapter on memory, I explained that it is essential we exercise our cognitive control system during our school years, because it trains facilities that underlie all thinking. Now we see it also offers fundamental tools for emotional control. Brain observation reveals a principle that surpasses our intuition: exercising our working memory and attention is an excellent way to take care of our emotions.

The story gets even more interesting. The brain circuits involved in resignification are those in the prefrontal cortex, which are responsible for maintaining and redistributing information. In this case, their role is to connect emotional information in the amygdala with the brain's language areas. The regions that monitor and inhibit other brain processes, such as the anterior cingulate cortex, are not involved. In other words, resignification is not based on inhibiting brain processes, but on redistributing them. This makes it much more effective, because the inhibition of brain processes often causes the opposite effect, as demonstrated by the famous pink elephant experiment.

In this cognitive control classic, people are asked *not* to think of a pink elephant, or something else that no one would ever think of. I usually do this experiment at talks, asking the audience to applaud every time they think (to their regret) of a pink elephant. The same thing always happens: one person claps, then a few, and by the end there is vigorous applause. Here's the irony: asking people not to think about something is the best way to get them thinking about it. Why does this happen?

The synapses of a neuron send *signals* according to the

neurotransmitter they release. A neuron that releases *glutamate* excites those with which it connects, and if it releases *gaba*, it inhibits them. In a neural circuit, excitation and inhibition overlap, forming a balance. When an inhibition process is triggered, *gaba* is ingested, breaking the equilibrium, causing an excitation rebound. This very schematic explanation illustrates the swings of neuronal modulation that cause instability. As a result, every time you try to inhibit a process, you also provoke it.

This becomes even more noticeable when inhibition is unconscious. Let's see how it works: cognitive control circuits function by stereotypical conventions. For example, when we see an arrow, our attention is automatically directed in that direction. We can also learn atypical conventions: for example, directing our attention in the opposite direction to the arrow, if it is red. Here two systems of the Posner network come into conflict: the exogenous, which carries attention in the direction of the arrow, and the endogenous, which inhibits this mechanism.

All of this happens in the conscious sphere. Playing this game with subliminal presentations has shown that cognitive control can also operate from the unconscious, but by its own rules. When subliminally presented with an arrow pointing to the right, our attention is directed to that portion of our visual field even though we haven't seen anything. However, when the red arrow is presented subliminally, we automatically bring our attention to where the arrow is pointing. In the subliminal indication, the red "no" stops working, and what was intended to be inhibition becomes excitation. That is why the paradox of control becomes even more evident in the sphere of the unconscious. *Trying to inhibit something is usually the best way to provoke it.* Although the connection is a bit remote, we can relate this idea to the Freudian notion of repression. The mere indication of inhibiting an idea (or an emotion) causes the opposite effect: it consolidates and continues to bounce off the neural confines of the brain.

3. **Brain activity during resignification is lateralized with dominance in the left hemisphere**[44]

The conceptualization of the brain into lateral

---

44 We see how the technical jargon spreads by repetition in the famous Marx Brothers' parody in *A Night at the Opera*:
"Now pay particular attention to this first section because it's the most important. It says, 'The party of the first part shall be known in this contract as the party of the first part.' How do you like that? That's pretty neat, eh?"
"No. It's no good."
"What's the matter with it?"
"I don't know, let's hear it again."
"Says, 'The party of the first part shall be known in this contract as the party of the first part.'"
"Sounds a little better this time."

hemispheres—one half rational and the other half emotional—is a metaphor that causes more confusion than clarity. Despite that caveat, it should be noted that there are specific cases of functional specialization by cerebral hemisphere. For example, in language, where the most decisive regions for understanding and articulation (Wernicke's and Broca's areas, respectively) are located very predominantly in the left hemisphere. Thus, the lateralization of a task is usually a sign that language plays a relevant role in its articulation. This is most surprising when it happens in cases where verbal storytelling is not explicit and language works from the unconscious to solve mental problems. The fact that the brain activity of a person reappraising an emotion is greater in the left hemisphere indicates, therefore, that this regulation process is built on language. Our look into the brain corroborates that language has an extraordinary ability to reappraise our emotional experience.

4. **During resignification, we see activation in the brain regions of the "theory of mind" system**
The studies I've used to explain the components of resignification are based on the observation of other people's emotions: participants observe images and try to distance themselves or interpret what they've observed more benignly. The simplest way is to assume that it's all just a performance. One possible objection to the results of these experiments is that it is much easier to reappraise the emotions of others than those we feel in our own body. True. But we can use this very circumstance to our advantage. During resignification, the brain circuits of a system known as *theory of mind* are activated, which allows us to attribute thoughts, feelings, and intentions to other people. This implies that a component of resignification involves relating one's own emotions to those experienced by others. In these experiments, as when watching a movie,

it is about internalizing others' emotions. Reversing this flow, by thinking about our own emotional experiences in the third person, as if they were happening to someone else, offers us another extraordinary tool for emotional regulation.

Exercise I: Ideas about emotions, what they are, why they overwhelm us, and how we can regulate them

Reminding ourselves that a dying character in a film scene is just acting is one of the simplest forms of resignification. Throughout this book we have seen more powerful ideas: reinterpreting fear as enthusiasm, sadness as anger, or rage as ecstasy. In each of these examples, understanding distances and proximities gives us a guideline for determining which emotions are most prone to resignification. This seems like a good time to offer some concise sketches of a few emotions: their nature, their motivations, why they sometimes turn against us, and clues as to how to regulate them. Here I intentionally depart from the rigors imposed by science and enter the realm of storytelling with some examples that don't claim to be common to everyone. I follow the path indicated by Leonard Cohen when he said about songwriting: "Your most particular answer will be your most universal one."

Fear

Palpitations, shallow breathing, sweating, tremors, high blood pressure, gastrointestinal upset: when we feel fear, our whole body reacts like an alarm. The threat is always clear and precise: a gun is pointing at us, an animal is growling at us, we can glimpse an impending cliff . . . That precision distinguishes fear from anxiety: with fear we know what we are threatened by.

Fear has a paradox: it directs us towards what we want to avoid. One of the first lessons those walking near a cliff are told is to keep their gaze straight ahead, avoiding the urge to look into the void. This natural reflex of paying attention to what scares us is, in this case, harmful: where our eyes go, our body follows. That's why it is important to take precautions. Fear is sometimes a hypersensitive alarm system that magnifies risks and makes us focus much more on them.

The reaction to fear is fight or flight, paralysis, or submission. None of these responses is appealing, but we can reconcile ourselves to them using the following perspective: fear is our body warning us to take care of ourselves. We can embrace fear the way we embrace a friend who looks out for our well-being.

Repulsion

We identify disgust almost instantly by its manifestations: wrinkling our noses, backing away, feeling sick to our stomachs. Repulsion, like fear, wants to protect us. Just as we can respond to fear by fleeing, disgust prevents our bodies from getting close to repulsive things and even expelling them if necessary. It defines a perimeter to mark repugnant objects as trespassers.

Disgust causes us to reject some things sensibly, such as toxic mushrooms, but sometimes it also leads us to reject other things that are actually good for us. Sometimes repulsion mixes with morality. You can feel disgust towards someone for how they eat or how they look. That repulsion no longer merely excludes repugnant objects, but leads to rejecting people. Repulsion makes us intolerant. More generally, it thwarts curiosity. It prevents us from exploring things, people, and worlds because it censors them hastily.

There seems to be a clear line separating what disgusts us from what doesn't. Nothing could be further from the truth: that border changes constantly. First of all, it changes over time. As children we find coffee, strong tea, beer, and most bitter tastes disgusting. (Think of a child taking their first sip of beer, and how evocative of disgust their facial expression is.) Yet over time we can come to enjoy those flavors very much. Second, we don't find something disgusting if we consider it familiar or our own. Here the good examples are scatological. We aren't usually disgusted by our own farts, only by others'. Same smell, different interpretation. And it is a common habit, although embarrassing to admit, to look into the toilet to examine our excrement, which ceases to be disgusting if it is our own. Why is that? Third, there's the context. The common example is rubbing up against something or licking foreign fluids. In the heat of an erotic scene, we can perceive as pleasurable something that, outside of that particular context, would be frankly disgusting.

Repulsion is not absolute, it has many gradations, and we can reinterpret it so that it is fully expressed when we deem it appropriate and turn it off when it leads us to reject something that shouldn't be rejected.

Surprise

Surprise is a short-lived reaction to the unexpected. It startles our eyes open and stretches our body, as if we were making room for the new information we have suddenly acquired. Surprises are sometimes pleasant and sometimes sad, but in both cases they act in a similar way. They multiply subsequent emotions: sadness and joy are greater when they catch us by surprise. Surprise warns us that something has escaped our notice and expends enormous energy to prepare us against that happening again.

Surprise fails utterly in situations that are inevitably uncertain. Trying to control it is aspiring to control the world, which is not only impossible but also undesirable. A world in which we knew everything that was going to happen would be a terrific bore.

We can observe these two sides of surprise. We can choose to befriend astonishment, to spark our fascination and interest. We can appreciate its role in curiosity and the pleasure of discovery, and enjoy the many things that will surprise us. And, at the same time, we can ignore surprise when it is a startling warning that something unpredictable escaped our notice. There is no need for us to feel like idiots when we stumble over bad luck, nor should we feel like geniuses when fortune favors us. In both cases, surprise makes us feel responsible for a universe over which we have no control.

Sadness

While most emotions tend to set us in motion (hence the etymology of the word), sadness is the exception. It encourages us to stop, to take time to recharge. That's why it feels like a state of despondency. Its most eloquent expression is crying, whether heart-rending or silently contained. Crying is a way of communicating a state of fragility and our need for pampering and protection.

Sadness is usually born of a loss. But this loss is viewed from a passive point of view: we feel that we can do nothing to repair it except seek comfort. The most recognizable example of loss is the death of a loved one. Sadness is the way to openly express that we are disconsolate.

We respond in a more energetic and reactive way to a loss when we see it as unfair. In those cases, we tend to reappraise sadness as anger. The origin of the emotion is similar, but its interpretation and consequences are very different: it propels us to act to remedy the loss. Instead of accepting it and calling for some "down time," we spring into action. There is a fine line between anger and sadness. Sometimes, in the face of a loss, it is best to stop for a moment of reflection and ask yourself: am I angry or am I sad?

Joy

Joy is the emotion that makes us feel the most alive, it's the feeling that no one ever wants to fade. But, of course, that isn't possible. Take for example a celebration, one of the most paradigmatic manifestations of joy; it demands energy, it's distracting, and it ruins our ability to concentrate. A gleeful person is usually imprecise. This concept is much better understood in the third person. Imagine that we are on a stretcher about to undergo an operation and a surgeon comes in laughing his head off, in a fit of mirth. We would all prefer a serious one, focused on his surgical instruments and on our body, not on his exultation.

Here is the paradox: joy distracts us from the effort that led us to it. Great athletes learn to contain their celebrations because they know that victory isn't the end of the story, and that they have to keep training in order to be able to celebrate again. But there's more. Someone who's just graduated school or finished writing a book or reached a mountain summit usually stops for a second to think, amid their joy, of all the moments of fear, sadness, and frustration that led to that goal. And that makes them cry. It's a way to extend joy backwards in time, because while it's a beautiful feeling, it becomes even more beautiful with contrast and accrual.

Contrasting colors make life more interesting. That's why people say they wouldn't choose, if it were possible, to live in a world of only joy. A full life with many interwoven emotional nuances is often preferred. This suggests another powerful tool of control: the next time fear, sadness, or anger appears, we can try to squint to see the joy hiding there in the distance, to understand that there is no sorrow without laughter or laughter without sorrow.

Anger

This emotion is born of an unfair offense. Anger energizes us, motivates us, organizes us. So much so that, without realizing it, we invoke it. Before a fight, an argument, or a match, we usually trigger anger by remembering a painful defeat or a comment that irked us. It is a way to prepare our body for battle. Anger has a positive side: it makes us see and express what we believe is right. Thus, anger reveals our values and bonds us to other people who share them. It creates identity and even pride; it unites and motivates.

Each of these values has a harmful version, especially when exaggerated: fights between siblings, couples, parents and children, or between two strangers because of a simple scratch on a car . . . Each of these situations originates an impulse that leads us to confuse that fight with "the big fight" and prepares us for a much more crucial battle than the one we are actually fighting. Anger leads to more anger, and draws our focus to arguments that trigger anger. And then everything explodes.

Distance, whether geographical or temporal or in perspective, mitigates this emotion. We become enraged in the heat of the moment by things that the next day, or when told in the third person, seem ridiculous. This reveals an important lesson: it is always good to take some distance, wait a second or two or ten, step back, or even, although it seems risible, try to think about the problem in another language. This is actually a fabulous trick for those who are bilingual to try when arguing with their significant other: switch languages. The distance and perspective provided by the other language make it much less likely that our anger will lead to more anger.

Jealousy

Jealousy is the emotion of possession and property: we are jealous of what we consider belongs to us. On the other hand, what we want to possess, but don't, arouses our envy. Jealousy usually surfaces in romantic love, although it is not its first manifestation. As children we feel it when a new sibling arrives, and we have to share our toys and our parents. The threat of our things being taken away from us has been with us throughout our lives. Why? Because we build our sense of self with our belongings. My toys, my parents, my siblings, my partner: they are all building blocks in the narrative that shapes our identity.

Being jealous is an automatic response to taking care of our most immediate world. But when we take a closer look, we discover that it isn't the only or even the most suitable response, and we also realize that there is something liberating about accepting the eventual loss of our intimate sphere. It brings the same relief as traveling light. Jealousy, like the fear of heights, seems innate and somehow irremediable. But it is neither of those things. Jealousy is learned and constantly being redefined; it varies greatly from one person to another and depends on who they are relating to. There are some people who go through several relationships without the slightest hint of jealousy, and then suddenly it rears its head powerfully. And it settles in, like a virus that's very difficult to shake.

This emotion leads us to restrict the life of the person who provokes it. It usually comes up around sex, as it manifests more easily in the realm of romantic relationships. But we can also be jealous of their friends, their dance partners, their privacy. Would we feel jealous if our romantic partner slept spooning someone else? If she were hugged by a friend in greeting? If he walked hand in hand with someone else? Each of these questions defines the boundary of what we consider ours, of what we believe we have the right to forbid in our partners' lives. But they also serve to conjure up the opposite image: in general, we hope that our loved ones will receive hugs, walk hand in hand with others, laugh, have friends, and enjoy themselves. There are many ways to love and everyone finds their own. But it is good to examine the different options and understand that the very thing that makes us jealous can also lead us to enjoyment. These reactions seem diametrically opposed, but they are much closer than it appears. There is a universal experiment for jealousy: what are the people who snoop on their partner's phone or Instagram expecting to find? It looks like a police background check, but it is also, partly, literary research. They are looking for something that serves as fuel for the narrative of jealousy. Is that worth doing?

Exercise II: Ideas for living better

1. **Avoid loneliness**
   Loneliness is having no one to talk to, no one with whom you can share joys, sorrows, worries, and successes. The effects of loneliness are powerful and severe: for both your physical and mental health. Talking not only helps us to reason and reach optimal solutions, it also helps us to take care of ourselves, have better control of our emotions, and prevent disorders such as depression, anxiety, and dementia.

2. **Physical presence also matters, especially in the worst moments**
   Being available to others and having others on hand when facing a difficult situation or serious illness is also an essential part of emotional care. These are the moments when it is most difficult to be by someone's side, but also when it is most necessary.

3. **Control is freedom**
   It is an old idea, but as valid as ever. Maintaining control in situations of high intensity, or great difficulty, is always desirable. Learning to control emotions, summon those we desire, and mold the ones that plague us before they take hold is a basic tool for living a good emotional life.

4. **Distraction is helpful, but often insufficient**
   The intuitive response to a negative emotion is to shift our attention elsewhere. It is a useful strategy, but it is only applicable in insignificant situations and has limited effectiveness: it does not work when facing great sorrows. Some distractions can also be more harmful than what they are intended to sidestep.

5. **You can induce emotions**
   With words, with gestures, with personal mantras. While not infallible, these inductions are sometimes enough to achieve the desired effect (to concentrate, clear your mind, or relax, for example).

6. **You interpret what you feel**
   The same sensations can have different meanings: the ones you imbue them with. Fear can turn into excitement, disappointment into acceptance, sadness into hope, etc. Learn to reappraise the emotions that hurt, obsess, or exhaust you. If you don't want them, turn them into something else.

7. **Reappraising does not mean suffocating an emotion**
   It means taking control of our own emotions, their details and nuances, and knowing the tools we can use to mold them according to our needs. Anger and sadness are as necessary as happiness, surprise, and love. Often, what we want to avoid is not the emotion itself, but the way it takes over our entire emotional space.

8. **Share your time with people who laugh**
   Laughter is contagious and a healing experience. People who laugh a lot tend to have fuller (and longer) lives. It is a catalyst for lasting personal relationships and an antidote to stress.

9. **Suppressing emotions doesn't work**
   Besides being a strategy that rarely succeeds in practice (try not to think of a white bear), it involves a high physiological cost in the form of stress. The intensity of an emotion is in direct proportion to how hard it will be for us to quell, and how high a price we will have to pay for that suppression. Perhaps it goes against our intuition, but it is always better to accept an emotion (and reappraise it, if necessary) than to reject or ignore it.

CHAPTER 6

Learning to Talk to Ourselves

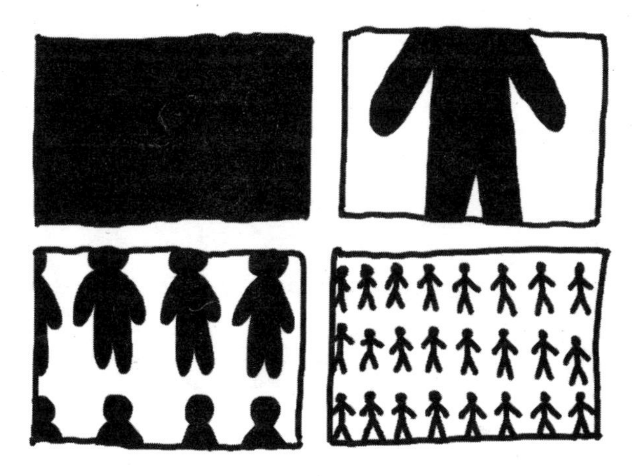

How to be kinder to the people we love most

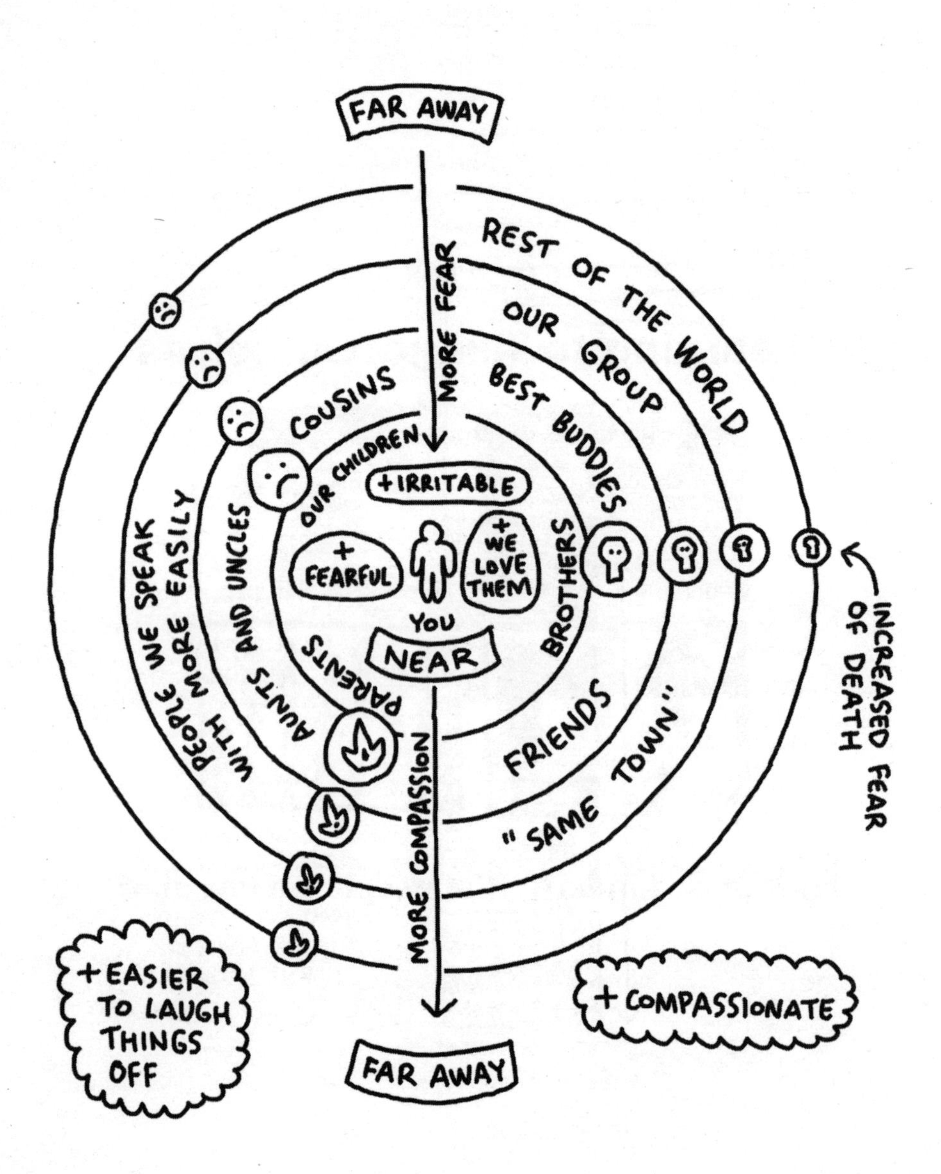

FAR AWAY
MORE FEAR
REST OF THE WORLD
OUR GROUP
BEST BUDDIES
COUSINS
OUR CHILDREN
+IRRITABLE
+ FEARFUL
YOU
+ WE LOVE THEM
BROTHERS
NEAR
PARENTS
AUNTS AND UNCLES
PEOPLE WE SPEAK WITH MORE EASILY
FRIENDS
"SAME TOWN"
MORE COMPASSION
INCREASED FEAR OF DEATH
+EASIER TO LAUGH THINGS OFF
+ COMPASSIONATE
FAR AWAY

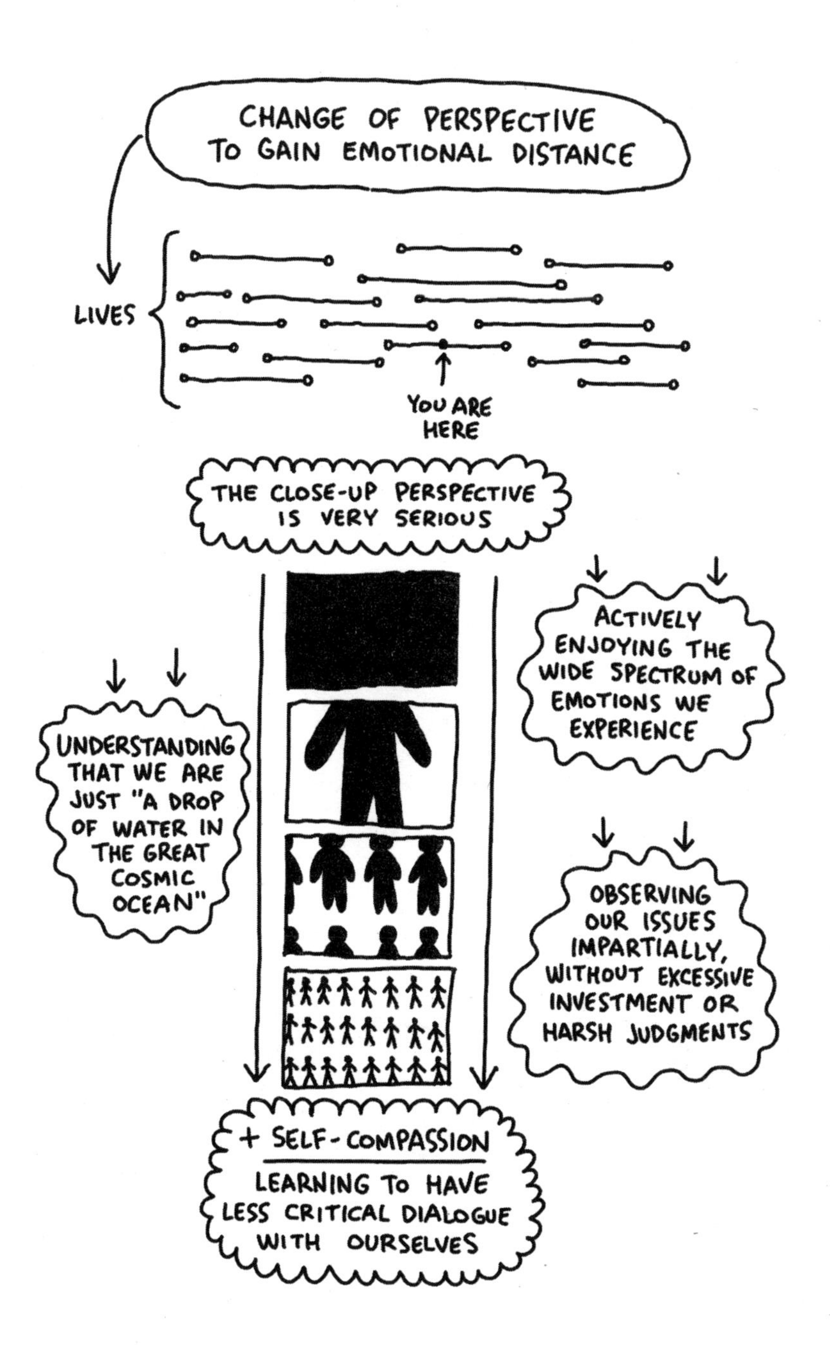
CHANGE OF PERSPECTIVE
TO GAIN EMOTIONAL DISTANCE
LIVES
YOU ARE HERE
THE CLOSE-UP PERSPECTIVE IS VERY SERIOUS
UNDERSTANDING THAT WE ARE JUST "A DROP OF WATER IN THE GREAT COSMIC OCEAN"
ACTIVELY ENJOYING THE WIDE SPECTRUM OF EMOTIONS WE EXPERIENCE
OBSERVING OUR ISSUES IMPARTIALLY, WITHOUT EXCESSIVE INVESTMENT OR HARSH JUDGMENTS
+ SELF-COMPASSION
LEARNING TO HAVE LESS CRITICAL DIALOGUE WITH OURSELVES

The closer we are to the people involved in an issue, the more our emotional sphere multiplies: we feel more love, more jealousy, more fear. We become particularly sensitive, and trivial matters irritate us more than they should. Things seem much less serious from a distance and it's easier to laugh at them. That's why the arguments couples have often seem ridiculous when seen from afar.

In the dead center of what's closest to us is, well, us. That's where things get really weird. We overestimate our achievements, but on many other occasions we are our most critical and severe judges. It is very difficult to see oneself in an even-handed and restrained way, without exaggerating the expectations, fears, and zeal that our experience of our own emotions provokes in our bodies.

When a stranger falls on the street, we make sure to ask him if he's OK. No one starts yelling at him, accusing him of having his head in the clouds. So why do so many of us react with a similarly critical lecture when it is our child who stumbles and falls? And when it comes to ourselves it gets even worse: we tend to blurt out a "How could I have been so stupid!" instead of giving ourselves a hug and asking ourselves: "Am I OK?" Why are we harshest with the people we love the most?

It turns out that both of these ways of reacting are the extremes on an axis that goes from compassion to its opposite: critical judgment. They are two ways of looking at and interpreting reality. The critical response is exacerbated by the wearying repetitions. Unlike the stumbles of others, when someone close to us falls it could create a new problem for us. Part of this unchecked reaction also comes from a desire to educate. Science teaches us, however, that this way of reacting is counterproductive. Getting angry with those who are already having a hard time is never a good solution. The compassionate gaze is much more helpful.

We need to identify the toxic habits that contaminate our speech and see how to engage in good conversations with ourselves. Science reveals how learning to be compassionate to ourselves helps us enjoy a better life.

Helen Keller was born in Alabama in 1880. Shortly before her second birthday, meningitis left her mute and blind. From then on her life becomes unbearable, as she has no way to express her desires and frustrations. There are no echoes or voices inside her mind; she cannot even hear her own screams. How do you learn language without seeing or hearing? Her parents ask for help from Alexander Graham Bell—who, in addition to inventing the telephone, dedicated a good part of his life to perfecting communication systems for the deaf—and, thanks to him, Helen meets the woman who will become her lifelong companion: Anne Sullivan.

Anne looks for ways to introduce language through touch. For example, by repeatedly moving her fingers on Helen's palm, as if she were writing. She patiently persists, making no progress, until, one morning, she spells out the word *water* on her student's hand and then takes her to touch the water. She repeats this process until, at one precise moment, it becomes evident in Helen's expression that the miracle of language has just happened: she understands that this drawing on her palm is a way of referring to water that she can use to ask for some whenever she is thirsty; she discovers the symbol and the word and realizes that they will be the key to open up her mental prison.

Emboldened by her progress, Anne ups the ante. She places Helen's hand at her throat as she speaks slowly to her. Helen can't hear Anne's voice, but she can *touch* it, she can feel the vibrations of someone else's vocal cords in her hands instead of her eardrums. And, with extraordinary effort, she learns to reconstruct sound by touch, and is able to bring her own voice to life.

Perspectives near and far

A friend who lives in Switzerland told me how one day her son ran out into the street. Obviously it scared her half to death, and she responded by shouting at him. She then told me that someone who was walking by admonished her and later reported her for having raised her voice to a child. The matter is complicated, of course, like all ethical and moral issues. Is there ever any justification for yelling at a child? Many people think that some yelling is a necessary and fundamental part of raising a child; others are convinced it is never justified.

I have no interest in moralizing; there are plenty of other books and other thinkers who do that. My intention, rather, is to break down the elements that are put into play when we think about them and turn them into emotions. My friend's anecdote is useful as a caricature of something that happens very often: a child takes off running, falls, and starts crying; her parents spring into action, with two simultaneous impulses: one is to hug her and the other is to scold her.

Some parents are compassionate and choose to comfort her. Others are more critical and prefer, instead, to rebuke and instruct. Science, however, shows that compassion is much more effective than criticism. The moment of an accident is not the ideal time to yell or try to teach: it is when we should give our kids a hug. There will be time later for the calm explanation of strategies so that something like this does not happen again. Only then will they be truly effective.[45]

Whether someone chooses to be critical or compassionate depends, to a large extent, on their personalities. But it depends even more on who is in charge. We are usually tougher on the people we love the most. Siblings, parents and children, couples who stop talking to each other for life. The anger that results from dysregulated emotions—jealousy, frustration, envy, and, above all, rage—can be attenuated with a more compassionate posture. So how can we change our perspective? We will see the solution in more detail over the next few pages, but let's begin with three very simple principles. First, it is almost impossible to judge someone without walking in their shoes, even when we live in the same house. Second, the main axis that separates the compassionate gaze from the critical one is the affective distance. As such, a simple way to be more even-handed and compassionate is to take a step back. We will see how humor, among other tools, helps us to gain some distance. Third, kindness and generosity are incredibly reflexive and contagious: being kind to others is the simplest way to be kind to yourself.

Because, at the end of the day, the person we are least compassionate towards is ourselves. When we stumble, we rarely console

45 In the end, as per usual, the Swiss are right.

ourselves. It's much more common for us to flog ourselves and think we're idiots: "How could I have made this mistake?" In general we are quite bad at speaking to ourselves and this affects the decisions we make, the way we shape our memories, and our emotional life. Once again, the recipe we've already seen in each realm of thought reappears: talking with others in order to hear their opinions and, above all, to express aloud ours; talking with others in order to learn to talk to ourselves, to review, organize, and shape our ideas. In short, talking with others in order to learn how to think.

When it comes to taking care of our health, we tend to be more compassionate to others than we are to ourselves. The advice we give to someone feeling poorly is very different if it's in the first or second person. It is very common for us to recommend a sick friend visit a doctor. However, our advice to ourselves is based on a very different scale of values: we have no time to take care of ourselves because there are "more urgent things" to be done. Sometimes we even brag about ignoring symptoms or possible health risks. Paradoxically, this often occurs among doctors, who do everything in their power to heal their patients, yet neglect themselves.

The weight of our own magnifying gaze

Obsessive self-judgment is a result of a lack of perspective that leads us to overestimate our problems. Meditation can help us to gain perspective, so we can understand (and feel) that our own experience is merely a tiny portion of a very vast universe. This distance lightens our experience, diluting it in a wider cosmos.

We all tend to overestimate the relevance of our work. There are many demonstrations of this concept; my favorite is the one my colleague Dan Ariely did, by bringing together a group of people to make origami, including some who were experts in this art. Then he asked another group of people how much they would pay for the various paper figures. Origami is a complex art and it is not easy to put a price on it, but people aren't stupid and they can see quality. As a result, the origami made by the experts were priced higher than those made by the amateurs. The amateur participants also understood the general rule—the origami made by the professionals were worth more—but everyone was convinced that their work (and only theirs!) was the exception that confirmed the rule.[46]

There is another similar example: the photos of our newborn babies. Everyone thinks their baby is the cutest one in the world. Of course, they can't all be objectively correct. But for each parent, it is true: we see our own creations as more luminous and transcendent. There is something beautiful about this argument, especially in these examples that brim with love for the people and things that surround us.

A few years ago I devoted a lot of my free time to studying music and soon after I composed my first melody. It sounded so beautiful to my ears. I had the feeling that, if I had to sum up the history of music, the three milestones would be Mozart, The Beatles and . . . my song.

46 It is difficult to have an objective perspective of one's own achievements. As seen in the joke about the two chickens walking together; when one of them says, "Hey, what's up?" the other one stops in his tracks, shouting, "Chickens can talk?!"

Here I was able to witness firsthand how powerful this illusion is, although it does not show up throughout the entire creative sphere. In writing or science, my usual creative spheres, the critic in me emerges much more strongly than the enthusiast. Yet with music the illusion was tremendously powerful. It was an example of the amphibious condition between fact and fiction: I understood perfectly that my melody was *just* a composition by a beginner. But this did not alter the illusion in any way, which motivated me to continue practicing music, something I found enormously challenging. This, I think, is a reasonable way to live with our illusions: by giving them an outlet to encourage us to be more intrepid, without losing sight of the fact that it is just a good illusion.

In the *real* version of Ariely's experiment, we tend to forget that our own work is magnified by a sense of excessive relevance. It happens to the retailer who loses a sale and experiences it as a catastrophe. And of course it also happens in art, science, and sport, where our vanity is at stake. I remember the frenzied and ruthless work climate at university when I was completing my PhD. The pressure sometimes was a consequence of our own passion as researchers, but it mostly came from our perception that our slightest miscalculations would be a huge disaster. We each experienced our own experiment as if it were *the pillar of the world*. The laboratory lights were on throughout the night, and the students worked nonstop from Monday to Sunday. When we saw that behavior in our classmates, we understood that it was disproportionate: the world wouldn't stop spinning if they didn't write that article, science wouldn't disappear, and even their academic career would be largely unchanged. Yet all that was instantly forgotten when we returned to our own reality.

This is when the *illusion* that magnifies the value of what we do stops being a healthy motivation and becomes harmful. It can break up couples and families, and lead us to neglect our physical bodies, and lose our sense of humor. And, in some cases, it can even end in tragedy, like what happened to Jason Altom, a student of Nobel laureate Elias Corey, who committed suicide by drinking potassium cyanide. Before he died he wrote a note in which he accused his director of having depleted and exhausted him. Corey responded by saying that he was devastated and that the pressure they had put on themselves to achieve such an extraordinarily complex chemical synthesis was mutual and shared. Be that as it may, what is clear is that the stress was brutal and based on the premise that there was nothing more important in life than their scientific project, not even taking care of their physical and mental health. Of course, it isn't easy to find the right balance, because that same ambition and borderline insane work ethic have been the fuel of many human accomplishments that we applaud unreservedly.[47]

In recent years, this tension has really shown up in highly competitive sports. We have the famous case of Simone Biles, a top gymnastics star, who left the Tokyo Olympics to take care of her mental health. These are exceptional examples, yet they illustrate something that happens to many people. Ultimately, it has to do with the meaning we give to the word *success*. With the critical perspective that we usually apply to ourselves, success is associated with professional goals: achieving a certain number of sales, contracts, medals, or followers on Instagram. This reference is always normalized according to its own scale. Obsession, fanaticism, and excessive demands are expressed as often at Little League games as they are at the board meetings of large corporations.

On the other hand, when viewed with the compassionate perspective we generally apply to our friends and our relationships based on a simple, kind (and not possessive) affection, our notion of success is very different. We don't love our friends more because they've sold more cars, closed a more profitable deal, or stitched a patient's wound a little better. We love them because we have fun with them, because we can hug them, because they're there for us when we need them, and we're there when they need us. This is a very different idea of success.[48] Remember this from time to time and treat yourself as you would treat a friend.

---

47 As we saw earlier, Aristotle noted that we have no word for moderate ambition. Long after his death, it's still hard to find it. How to maintain ambition and the fire of desire without neglecting life's other dreams?

48 The vague categories around words are in this case quite eloquent. The term

The clearest examples of perspective distortion are found in our closest relationships. The family dinner table is a good case in point. Parents are usually quick to notice (with their magnifying gaze) if their child eats with her hand, or with her mouth open. We've all seen extreme cases of a parent who insists he must continue edifying his fifty-year-old child on the basics of life. The bond with a niece or nephew is usually very different; because there is more distance and less responsibility, the relationship has a looser, more playful tone. The same things that make us angry in a daughter, we find amusing in a niece. But it is still an educational bond. We can learn essential naughtiness from our aunts and uncles, something that our parents are unlikely to teach us. I suggest a simple exercise to flex our perspective-changing muscles, and help us to understand the effects. Try relating once a week to your children as if they were your sibling's children. And see what happens.[49]

---

success comes from the Latin *successus* which means "result" or "outcome." Success is usually the applause at the end, which is stimulating and magnetic but can leave us with a gaping void because it forces us to abandon what led us to that celebration.

49 Summon the amphibious nature that allows us to flirt with fiction, and gain perspective to understand our relationships better. Let's be frog mothers and fathers, toad uncles.

The experience machine

This entire discussion revolves around a question as simple as it is unfathomable: what makes us happy? Happiness is a complex conglomerate; it is difficult to define and, as such, to measure. In some scientific studies people are simply asked to express, in numbers or words, their happiness. Yet we already know that we are not good judges of our own experience. Other studies measure happiness through laughter or the absence of stress. The truth is, however, that happiness involves a tangle of sensations, behaviors, and bodily states that are impossible to reduce to a single scale. As a result, many scientists and philosophers argue that a life with fears, anxieties, and love can be fuller than one of unbroken happiness.

The dilemma, of course, is an ancient one. Nearly two and a half millennia ago, the Greek philosopher Epicurus outlined a possible solution to the question of the origin of happiness in what he called *rational hedonism*. In his vision of ethics—that is, in his manual for a virtuous life—Epicurus begins by suggesting the obvious: one must seek pleasure and avoid pain. The question, however, is complicated when he adds that the pursuit of pleasure must be carried out in a rational way to avoid excess, not as a matter of morality, but because that causes further suffering. And this leads us to a real muddle, the sticking point that has stumped the whole modern science of happiness: how do we measure pleasure over a lifetime? Do we focus on its highest peaks? On the absence of moments of unhappiness? On an average?

Epicurus proposed a solution that was later taken up in the twentieth century by the philosopher Robert Nozick: happiness (whatever that is) cannot be reduced to a mere succession of pleasurable experiences. Nozick illustrates this idea with a dilemma that then inspired the red pill and the blue pill in the movie *The Matrix*: if we could connect to a machine that guarantees us that all our experiences will be pleasurable, would we? Nozick assumed that people would prefer not to, because they seek a link to reality, in addition to and above pleasure. Part of reality is experiencing a wide spectrum of emotions from various combinations.

Nozick's experiment has been done many times and, as he surmised, the vast majority choose not to connect to the experience machine. This does not prove any philosophy, but it does show that the intuition of Epicurus's rational hedonism about what constitutes a good life chimes with what people express twenty centuries later. It may not be universal, but it seems like a fairly well-preserved trait throughout the human condition. What is the most effective way to achieve this happiness that is more real and more colorful than a mere succession of pleasurable experiences?

The path leads us to ideas we've already visited: mitigating judgment, being kind to yourself, gaining perspective, reappraising "success," not feeding vanity. Connecting with real elements of the affective experience, with the people who are truly there when we need them, instead of with a horde of TikTok followers. Remembering every so often that we are merely part of a vast universe of matter, of galaxies of stars from whose dust sentient life is formed: this is what Epicurus's disciples must do, those of us who prefer not to be hooked up to Nozick's machine, to not take the blue pill.

Dying of laughter

Let's recap. The disproportionate focus we put on the people we love most triggers huge fears. We can alleviate them by gaining perspective, and also by recognizing fear as one of the many nuances of a life that is much richer and more real than Nozick's machine. Perhaps the ultimate challenge to this concept is the fear of death. Like the fear of heights, spiders, or snakes, the fear of death seems to be universally etched into our insides. But what if it isn't? What if it was a cultural invention or a result of the perspective from which we regulate our emotional experience? I specifically choose this fear, which seems almost impossible to transform, in order to challenge emotional regulation to the maximum.

We have seen how grief is reappraised through *liget*. Renato Rosaldo discovered that emotion among the Bugkalot, who experience death very differently than we do, with a hint of euphoria, sending a very high voltage circulating through the body. To reconcile us with this nonintuitive idea, Rosaldo reminds us that it is only the death of someone close to us that brings on *liget*; a distant death is experienced with indifference, even humor. That is why it is so healing to perform the reverse exercise: to take some perspective and understand that, as Jorge Drexler sings, "We are nothing more than a fistful of sea." "Calm / Everything is calm / Let the kiss last / Let time heal / Let your soul / be the same age / as the sky."

There is the possibility, at least as a mental exercise, of thinking of death as just another spark in the immensity of the cosmos. We are able to do that with other endings. Think of a wonderful opera, concert, or whatever form of human expression moves you the most. Now imagine it will never be performed again and that we've enjoyed that moment as one of the most fabulous of our lives. When the performance ends, we will feel nostalgia. But undoubtedly something much more like *liget* will also emerge: a high-voltage spark, a standing ovation, a burst of euphoria. Perhaps it will take the form of heart-broken crying, but with energy and the hint of a smile. Its last moments are the high point of the show, a celebration, as we see with contemporary works of art that are created to be fleeting so that their existence is more memorable. Perhaps it's possible to bring the experience of death closer to that feeling: a standing ovation, an applause for life. I know it seems impossible. To me, at least, it feels almost unattainable, like stretching a muscle beyond its limit. Yet perhaps it can serve as an idea to start exercising emotional regulation in kinder, gentler spheres?

There are also examples in Western culture of how to reassign the emotions around death into a celebration. One of my favorites is at the funeral of Graham Chapman, one of the members of the famous British comedy group Monty Python, during his friend John Cleese's eulogy, when he says that Chapman asked him to be "the first person ever, at a British memorial service, to say 'fuck.'" The people who are crying then start laughing. And soon the mood of the entire funeral service shifts; Chapman's loved ones bid him farewell with a remarkable collective paroxysm of laughter.

This is, by the way, humor's greatest virtue. It serves as an antidote to give us a bit of distance from uncomfortable, painful, offensive, or stressful topics. Laughter kickstarts our brain's *nepenthes* factory. Many times, like at Chapman's funeral, laughter helps us overcome adversity as a group, together, at the same time. Robert Levenson

explored this idea in some curious experiments in which he studies how couples respond to stressful conversations. At certain points in the discussion, some people will start laughing. And it turns out that laughter, in addition to being a fabulous antidote to reduce stress, is also an excellent indication of which couples will last the longest. The family that laughs together stays together.

José Mourinho, the Portuguese soccer coach, is a public figure who has chosen to express an arrogant and conflicted personality; people either love him or hate him, with no middle ground. What's curious is that how Mourinho is received varies by location: in Spain he is mostly hated, while in the UK he is adored (or was for a long time, before a second much more difficult phase). Why does the same person arouse such different emotional responses? Sports journalist John Carlin has an edifying suggestion. In Spain Mourinho is taken seriously and in the UK he is taken lightly. As if the British see Mourinho as a spectacle, something humorous. He is not taken as seriously, they are able to distance themselves from his character. This is another useful resource for our tool kit, like the roller coaster: thinking of things as mere entertainment. Soccer is not that serious, Mourinho is not that serious, fights and arguments with our partner are almost never that serious; perhaps death isn't even that serious.

Hugging ourselves with words

Talking to ourselves in a more even-handed and compassionate way helps us enjoy a much better life. Self-compassion is a complex idea imported from Buddhism and, as such, is in curious dialogue with Western science, which is an intrinsically reductionist tradition. While curious, this meeting of East and West has been highly productive and has resulted in three decades of theoretical, experimental and practical research that we can each bring to our own lives.

We talk to ourselves all the time. All it takes is three seconds alone (without a cell phone) and voices in our head will start talking to us about what we have to do the next day, about whether a certain game could have ended differently, about some song by Nicki Minaj, about our romantic partner, about our ex, about the exam coming up, or about what we should have said in the conversation we had right before we got into the elevator. The brain has a default mode of functioning, which involves a set of regions distributed mostly along its medial line. This brain system alternates with another network that controls everything we do deliberately, with effort, with a specific purpose and focus. When our mind wanders as we read, or when it starts producing voices as we stroll, it is because our brain has gone into default mode. Daydreams are a substantial part of our lives; when they interrupt our thoughts, they disrupt other brain processes. The brain retreats to its own factory of ideas. Chris Frith, a great British neuroscientist who studies consciousness, argues that we all constantly produce mental delusions similar to those experienced by schizophrenics. The difference is that the healthy mind recognizes those inner voices as its own. Except, as we have already seen, when creativity is put into play.

The journey of inner voices

In recent years it has been pointed out that, in fact, the conversations we have with ourselves are usually toxic. Psychologist Dan Gilbert found that people tend to feel less happy when their mind wanders into inner conversations, because those voices are often fraught with anxiety and frustration. No one has ever taught us how to be travelers within our own minds, so when the brain is left to its own devices, it tends to converge repeatedly in obsessive places. Someone who has suffered with jealousy is likely to get caught up in that experience when he then establishes a new relationship. Someone who has felt a lot of anger will constantly see the world through the filter of rage. The mind has enormous inertia and, like babies who cannot stop crying of their own free will, it has a hard time leaving the regions where it has settled.

Stanislas Dehaene, an extraordinary neuroscientist and my mentor, has a world map in his house where he marks every place he's visited. Since Stan is fearless and curious, that map helps tip the scales every time he receives an invitation. If he's invited to a country he's never been to, he is much more tempted to accept. The journeys of our mind, unless we are determined to change the default mode of our brain, usually work with the reverse logic: we tend to return to the territories that are already marked, like people who always travel to the same places, except that our minds, left to their own devices, will never choose bucolic landscapes. They will usually journey to dark recesses.

So in order to learn to be compassionate with ourselves, we must first unlearn the extemporaneous way we tend to speak to ourselves. To shift the habit, tone, and style of our ruminations so that our conversations with ourselves are not pitched battles inside our own minds.

In 2003 Kristin Neff of the University of Texas published a paper entitled "The Development and Validation of a Scale to Measure Self-Compassion." It is a fine example of how to take a broad concept from another culture and mold it into an object of scientific study.

Self-compassion is measured by having participants indicate their degree of agreement with statements like: "I try to be understanding about the parts of my personality that I don't like," or "When times are really tough, I tend to be hard on myself." These two scenarios are related to the first dimension of self-compassion: the axis line that goes from a critical judgment to a compassionate one.

The second dimension measures the observation perspective, either with one's gaze focused on oneself or on the common human experience. Living life with the knowledge that we are, to paraphrase Drexler's song, "just a drop of water in the cosmic ocean." Two more scenarios on the questionnaire are: "When things go wrong, I see difficulties as a part of life, something that everyone goes through." And at the opposite pole: "When I get depressed, I usually feel that most people are happier than me."

The third dimension serves to determine if one approaches one's emotions with an open mind, understanding that feelings are usually complex and have many facets, or if, on the contrary, one focuses on specific, generally negative, aspects. The two scenarios that measure the poles of this predisposition are: "When I feel bad, I try to approach my feelings with curiosity and a receptive attitude," and "When something painful happens, I tend to obsess and fixate on everything that's wrong."

On the Neff scale there are a total of thirty scenarios, five for each extreme of the three dimensions. It is relatively easy to position yourself in each of these statements. In fact, they seem somewhat trivial. Yet this is the most direct and effective way to quantify self-compassion. The three elements that make up self-compassion are fairly independent of each other, and can show up to varying degrees in different people. A good mix of all three is what makes us self-compassionate.

Enumerating these "atoms" of compassion also allow us to clarify its definition, and not confuse it with empathy. Both involve an emotional bond with others, yet there is a substantial difference. Empathy means reproducing others' sadness with sadness of our own; compassion, on the other hand, involves the intention to resolve and remedy that sadness. It is also often confused with excessive self-esteem. But, unlike compassion, disproportionate self-regard has a narcissistic component. Those who are compassionate are even-handed. They don't love themselves too much, they are simply kind and have an open predisposition. We will later see that these two ways of loving ourselves, one with understanding and the other with adulation, produce substantial differences in the way we relate to our own experiences, failures, ideas, and emotions.

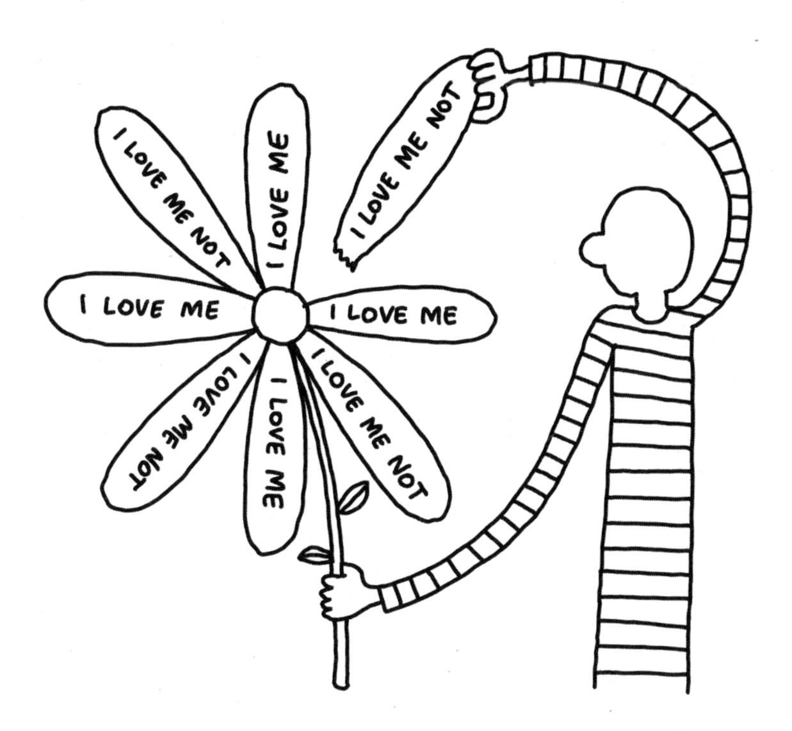

Compassionate science

Each person comes "factory-equipped" with their own baggage, some with a predisposition to be more critical and others to be more compassionate. We can change that predisposition, but it requires some practice, motivation, and adequate expectations. We can all make a little progress, and those small improvements can sometimes lead to substantial changes. It's worth a try, because, ultimately, we live everything through our mental experience. It's all we have.

The basic ingredients are willingness, time, and realistic expectations, and to that recipe we must add one more thing: a good method. And here we find that there are recipes for all tastes. Some key ideas are repeated in each one. First, the way to practice self-compassion is via compassion, which comes much more naturally. We've already discussed this formula: talking to others teaches you how to talk to yourself. (We have also mentioned the paradox that compassion becomes more challenging the more we love someone, the closer we are to them.) Second, like all mental or emotional regulation work, it is essentially an exercise of focus. You have to start directing your thoughts towards certain ideas, despite there being much more attractive options in the vast mental universe. It is about an efficient use of the cognitive control system, which we have already examined. Focusing on the desire for someone else to be fine—that's pretty simple. But often our minds wander and are contaminated by other ideas. A large part of the repetitive exercise entails sustaining the compassionate state. This takes time and effort, and progress is slow and gradual. There is a certain similarity with physical exercise, not physiologically, but as a metaphor. Our task is to awaken the compassionate state so that it becomes more prevalent in everyday moments of life.

Self-compassion works: it provides us with a better physical and mental life without making us self-indulgent. To prove this we will again turn to science. The studies on the evolution of AIDS patients demonstrated that loneliness exacerbates the deterioration of their health. We also saw that the opposite of loneliness is not having thousands or millions of followers on social networks, but having someone to talk to in a frank, warm, and understanding way. It is impossible to achieve this sort of conversation with a crowd.

We can give a precise name to the opposite of loneliness: it is compassion. And it is important that we are compassionate to the person with whom we talk the most: ourselves. As we will now see, self-compassionate people experience less physical and emotional toxicity.

Neff discovered this in an experiment that asked participants to speak in public and to solve mathematical problems, two typical sources of stress. Self-compassionate people pass these tests with better rates of resilience and psychological well-being. This change of perspective also affects the body, particularly the sympathetic nervous system, which prepares us to respond quickly and actively to danger. Its counterpart is the parasympathetic system, which places the body in a state of rest, relaxation, and energy-saving. More compassionate people secrete less alpha-amylases and interleukins, two molecules that increase dramatically when the sympathetic system overemphasizes its response to stress. In other words, a compassionate mindset keeps our bodies safe from their own over-reactions.

Another key study on the potential for self-compassion was done by Mark Leary's team, at Duke University. It is helpful for us to examine the layers of this experiment because they each specify situations where a lack of compassion can result in all kinds of toxicities.

In the first study, participants remember the worst thing that happened to them in the previous four days and indicate how much of each emotion that memory contains. It turns out that self-compassionate people label their memories with less negative emotions. This in turn changes the overall tone of our autobiographical narratives, the quality of how we see ourselves in the mirror.

Could it be that less self-compassionate people make their stories worse as they tell them, or is it that they have more negative experiences available in their memory? The second study answers this question by presenting all the participants with exactly the same fictional scenarios: in one they fail an important test; in another they are responsible for the failure of their team in a sports competition; and in the third they forget their script in a public speech. Just as before, self-compassionate people evaluate these experiences with less negativity, tending to consider them normal setbacks that are merely a part of life.

The third study is the one I find most ingenious. In it, participants are asked to introduce themselves and say what they like, what they identify with, their dreams, and their life plans. A few minutes later, their presentation is evaluated by another person. But, of course, there's a trick: the evaluator's opinion is manipulated at random; some participants indiscriminately receive praise while others receive criticism. The results reveal that self-compassionate people are protected by two distinct layers. They are both less influenced by the external gaze, even when receiving a randomly unjust negative assessment, and they are less likely to overreact to their own evaluation, even when it's not good.

This study helps us understand the difference between self-esteem and self-compassion: two gazes on oneself that, as we said earlier, look similar but are very different. The assessment by those with low self-compassion is worse than that of a neutral assessor. Their self-judgment is harsh and strict. People with high self-esteem, on the other hand, assess their performance much higher than is realistic. It's another distorted, narcissistic lens. Results show that self-compassionate people offer a calibrated opinion that is in keeping with the neutral evaluation. They treat themselves without flattery or contempt, in an even-handed and, moreover, kindly way.

This result explains the reason behind a mistaken intuition. There is a common belief that self-compassionate people do not improve because they don't challenge themselves, that if we aren't seen to be loudly chastising ourselves, well, then we aren't placing sufficient demands to thrive. But we see that self-compassion does not make us complacent. Quite the contrary: it offers us the opportunity to evaluate more objectively what we do. Without so much drama.

The value of a hug

I was traveling from Europe to South America. When I arrived at the airport there were two lines for the check-in. I joined the shorter one, where there was only one gentleman with a much younger woman. But such was my luck that the check-in process for this pair dragged on endlessly. At some point I approached them angrily and said that if they kept questioning every step, we would never get out of there. Later we crossed paths again in line to board the plane. The same three people in the same order. The gentleman, who now seemed much older than I remembered, said wryly to his daughter: "Let this man get ahead of us, he's obviously in a hurry." With some embarrassment, I said no. You go ahead. Walking through that tunnel on the way to the plane, feeling crushed by the fragile voice of that elderly person, my perspective changed completely. I understood that, consumed by anxiety, I had rushed a vulnerable person. And I decided to ask him for forgiveness; it was a negligible act, yet to me it seemed impossibly brave. I told him I was sorry, that the stress of the trip had made me lose my composure. The guy almost hugged me. I carried his suitcase, and walked with him to the plane, where I sat down and had one of the most restful transatlantic flights in memory. It wasn't because I had a particularly comfortable seat. If I hadn't had that change of attitude, I would have spent the whole journey mentally replaying the confrontation, analyzing who was in the wrong, thinking about how I'd had to wait, how much time was wasted . . . and I would never have managed to fall asleep. I would have spent the entire flight producing interleukins, scars, and resentment.

This is a very small example. I'd like to believe that all of us, at some point, have had the experience of a simple act where we reach out to another person, and everything changes for the better. Beyond wanting this to be so, I'd like to provide evidence that, indeed, the compassionate perspective does us good. Hedy Kober designed an experiment to prove this, using ordinary people who had no particular experience or aptitude for self-compassion, people like most of us. The participants were shown photographs with negative emotional connotations, and touched with a piece of hot metal, not burning them but verging on pain. The results were overwhelming: the brain's response to harmful stimuli changed drastically when people had a compassionate perspective; less activity is recorded in the amygdala and in brain areas associated with the perception of pain. And this happened without increasing activity in either the prefrontal cortex or other regions of cognitive control. The results were encouraging: it is possible *to cushion pain* with just words, without some superhuman effort to contain it, without gritting your teeth.

The most tragic and compelling demonstration of the power of compassion comes from Romania. Shortly after the death of Nicolae Ceaușescu, a network of orphanages was revealed, where some one hundred and seventy thousand children had grown up completely neglected. The first people to arrive at these orphanages, such as Nathan Fox from Harvard's Center on the Developing Child, say that there was absolute silence in the rooms. Without any affection or compassion, these children had lost their ability to speak. The babies were left in cots all day, completely abandoned until someone showed up to feed them or change their diapers. Not a single song was heard, not a single word; there was no kindness, no hugs. They lived in absolute neglect. It is the most tragic version of Cacioppo's experiment.

After the transition to democracy, many of these children were adopted and went through a number of studies. Almost all of them had severe cognitive problems. And their brains were abnormally small. If you look closely at the images below, you can also see that their ventricles, the cavities through which the cerebrospinal fluid flows, occupy more space and that there are fractures along the gray matter that denote cortical atrophy.

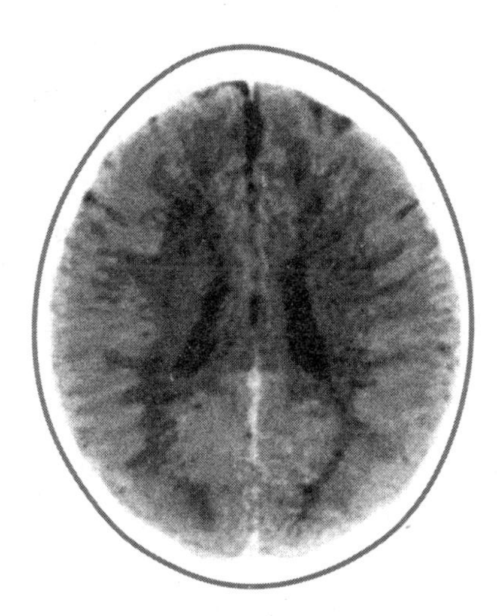

The brain of a child raised with normal affection and education

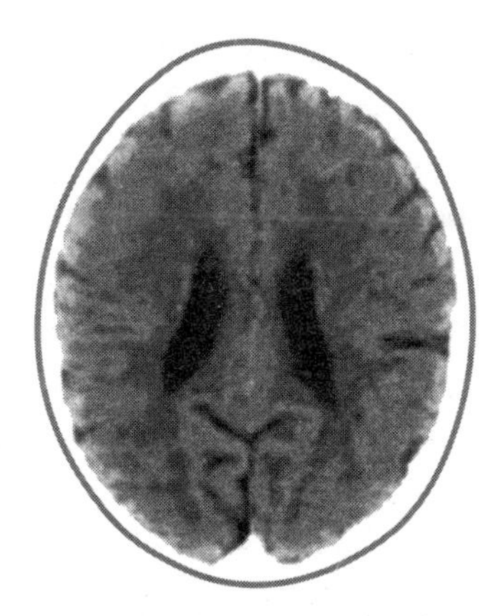

The brain of a child raised without affection, or educational and social support

This study shows us that affection is a vital fuel for normal brain development, and, at the same time, changes our intuition about how stress works. The most natural thing is to assume that stress results from an excess of toxic experiences, but now we see the opposite: that the absence of affection is a primary source of stress. That is why newborns, be they human or another species, seek out physical affection. That is the essence of compassion: offering relief in the form of gestures and words so that the brain and mind develop in the right environment.

On September 8, 1979, shortly before dinner, my parents left the house. It had been three years since the Argentine military coup that had made kidnapping its particular form of state terrorism. I was seven years old, and an ocean away in Barcelona, yet I had been poisoned with the latent fear that, at any moment, someone I loved could be "disappeared." When I woke the next morning and found my parents' bed empty, I felt that the hair holding the sword of Damocles had finally been cut. I was so blinded by fear that I forgot the most obvious explanation: my mother was nine months pregnant.

That day, she was at the hospital giving birth to Lucas. The terror of those years eclipsed even common sense.

A year later that fear—which so often struck at bedtime—vanished. Because before I fell asleep, I would curl up in my brother's bed and stroke his back as I told him stories. There is nothing more hypnotic than having your back rubbed. The fact that it is anatomically impossible for us to see or touch our own backs is what makes it so calming: we feel protected by another person, the one rubbing our backs. What's most surprising about that gesture is its powerful reflexivity, sending that peace and calm radiating back to the person who is doing the rubbing. Stroking Lucas's back every night at bedtime made my own fears evaporate and I was the one who ended up believing the stories I would tell him.

Exercise I: Practicing self-compassion

Among the many practices of self-compassion out there, I especially like the one Sam Harris included in his "Waking Up" project. Sam is a neuroscientist who's devoted an enormous amount of his time to meditative practice, merging these two passions into a project that forges a broad and open synergy between different cultural traditions. It has been very useful to me and has helped me shift my habit from a critical focus to a more compassionate one. This exercise falls into the larger category of mindfulness. It has a particular tone, like a secular ritual, and it is, above all, a way of conversing with yourself, slowly, rhythmically, with repetition, like the bedtime stories that were read to us as children. I realize that these sorts of mantras may rub some readers the wrong way, but I would encourage you to give it a try. In Harris's version, the practice of self-compassion consists of the following steps:

1. **Take control of your attention and bring it inside yourself**
   This usually starts by closing your eyes to avoid visual distractions. Then you have to focus on your breath. Of our many internal processes, breathing is the most tangible,[50] because it is felt in the abdomen and in the nose, and it can even be heard. In addition, our breath is rhythmic, and has a frequency that's not too fast and not too slow, which makes it an ideal hammock to help us turn our attention

50 There is also our heartbeat, although—as anyone who's ever tried to take their own pulse knows—it is more indirect than our breathing. What, you want another example? Well, try to perceive your pancreas secreting insulin.

inward. This idea has its correspondence in the brain. With Pablo Barttfeld we showed that the brain networks that light up when we bring our attention to our breath—or to our body more generally—are almost the same ones that bring our attention to our own ideas.

2. **Think of someone for whom you feel compassion**
   It is best not to think about yourself, or your spouse or partner, or your parents, or your siblings, because these relationships are complex, sometimes possessive, and provoke many conflicts. Ideally, we should think about a less intense relationship, about someone with whom we have a peaceful bond, someone who has never made us angry. The best person is a friend. In fact, the mere process of identifying this person is itself an excellent exercise.
3. **Bring your attention to the desire for that person to be happy**
   One way to achieve this is to slowly and sincerely repeat a series of desires. Like a litany or a mantra that has the power to transport your attention: may you find happiness; may you realize your dreams; may you live a life of peace; may you be safe from harm; may your friendships be deep. Then, in another attentional shift, try to turn your attention away from the person and onto the desire itself. Sometimes it helps to visualize yourself as a radiant source of light. And this brings us to the fourth step.
4. **Observe that this mental journey works**
   Turn a magnifying gaze on your own mind and observe the emotion we've induced with this exercise. It is similar to how our bodies feel after making love; recognize your heartbeat, your warmth, your perspiration, and any other expression that reflexively confirms the emotion we have produced. After bringing our focus onto compassion, we will feel that our mind is as flexible, relaxed, and calm as our legs are after we've stretched them. As simple as this

may seem, it is a small revelation. Compassion is also reflective. Wishing for good things for someone else softens and sweetens our own mind.

5. **Observe any possible distractions**
   This practice is not immune to the distractions that can take over our minds, as happens when we read or drive. The brain in default mode stages a mutiny in the midst of meditation. It is essential to identify this, contemplate it without judgment, and simply return to any of the earlier steps: our breathing, our identification of a person we feel compassion for, our wishing that person the best in life, or the reflective happiness this produces in us.

6. **Repeat these same steps with a closer bond**
The second stage is to practice compassion with a stranger, someone we pass in the street, the supermarket check-out worker, or the officer directing traffic. The key is that the fuel driving compassion is not the intensity of our love, but rather the absence of conflict. With this person in mind, repeat the steps: focus on your breathing, think of that person, understand how wonderful it would be if they were happy and how their happiness radiates outward and onto us, softening our minds and transforming our mouths, our faces, and our breathing. It makes us feel good.

7. **Bring your attention to the desire to alleviate suffering**
Once we've practiced the most accessible form of compassion, that of sympathizing with someone else's happiness, it is time to approach compassion's more complex, more relevant side: that of alleviating others' suffering. We again begin with a person for whom it is easy for us to feel compassion. But now that person is suffering; perhaps a loved one has died, or maybe that person is struggling with their own pain and imminent mortality. While the image evoked has changed, our desire for them has not. Herein lies the substantial difference between compassion and empathy. We do not cry. Our expression does not turn sad. Our intention remains the same. May you not suffer. May you feel peace. As difficult as it may be, we can concentrate and recognize the slight happiness this intention brings.

   We understand that the world is full of suffering. That people are born and die and that we all experience pain. That is not where we put our focus—not on our anger at the world's injustice—but rather on the love-filled intention we are radiating outward and on our profound desire for things to improve.

   The exercise is not easy, and it is slow going. Every so often our concentration goes to hell. Imagining a loved one sad can fill us with sadness. Then we must once again invoke our desire to alleviate that suffering. Every once in a while we will have to stop and return to the first step, to our breath.

8. **Direct compassion into the world**
   First we practiced compassion towards someone we have an easy friendship with. Then towards a stranger. Now we direct compassion out towards the entire world. As if we were Charles Xavier with his Cerebro machine, observing all the happiness and suffering of the X-Men. Think of all the births occurring in one second, all the parties, all the graduating seniors, all the first kisses. Think also of the wars, the diseases, the catastrophes, the poverty. In every instant thousands of millions of people laughing, shouting, crying, being born, dying. With this exercise we can eventually manage to radiate compassion out so generally that we can direct it at the entire planet. This serves to give us perspective. And that is necessary because in the next step we ourselves will be the object of our compassion, and the problem there is the space that we each occupy in our subjective representation of the cosmos.

9. **Self-compassion**
   We've practiced controlling our attention in order to focus within through our breathing. We've practiced projecting a sincerely compassionate intention, identifying it, seeing how it transforms us, endeavored to radiate a positive intention without being infected by the world's suffering. And we've practiced recognizing what infinitesimal specks we are in the universe. This path now leads us to move our focus onto ourselves, to make ourselves the object of our compassion. Turn it into self-compassion. And thus we will repeat those steps, wishing ourselves a happy, radiant life. Reciting the same intentions, now for ourselves, like a mantra: may I find happiness, may I realize my dreams, may I live in peace, may I be safe from life's trials and

tribulations, may my friendships run deep. And we repeat them, over and over. Because it is not enough to merely know something; the habit must be recorded through repetitions. We must listen to ourselves, wrapping ourselves in the warm embrace of sincere self-compassion.

Exercise II: Ideas for living better

1. **Give free rein to illusions**
   They are a good motivation, fueling your perseverance in unfamiliar or difficult territories. Celebrate the satisfactions of your successes and enthusiasms, without losing sight of the fact that they are often just that: an illusion.

2. **Treat yourself like a friend**
   We don't love our friends more because they've sold more items, worked more hours, or secured more "material" successes, but because we have fun with them, because we can count on them when we need them, and because they take care of us. Evaluate what things are truly important and judge yourself with loving kindness, the way you would a friend.

3. **Act compassionately**
   It's especially difficult to direct compassion towards yourself and your loved ones, but it's one of the most effective ways to suffer less and be happier. Be open-minded, physically affectionate, welcoming, accepting, and caring. Being compassionate to others is good practice for being able to transfer that same perspective to your closest friends and family and, of course, to yourself.

4. **Don't forget that improving your emotional life requires practice**
   Changing your emotional life means, to a large extent, changing the way you talk to yourself and ensuring your thoughts don't lead, by default, into dark places filled with

frustration, angst, and rage. This is not something that happens overnight and just wanting it is not enough. It requires practice, perseverance, hard work, and tenacity, because it means changing one of our most ingrained habits.

5. **Look for games to get some distance when in conversation with yourself**
   Some people write letters to their "future self." It's a good perspective-taking exercise, in which you write to another person who turns out to be the you yet to come. Reading these letters is also an interesting practice of dissociation from your identity: you read your own words as if they were written by someone else. What would you say to your future self? Usually, fears and anxieties emerge that you trust will heal over time, and you communicate them in the hope that your future self will have overcome them. Here is one of the many devices we have at our disposal for learning how to engage in good conversations with ourselves.

6. **Avoid the reflex of judgment**
   A meal, a painting, a person, an emotion . . . our first reflex is usually to judgment: Is it good or bad? Is it good for you, is it good for so-and-so? You can change this habit of judging every experience in life on a value scale. Experiencing an emotion without judgment gives you the opportunity to enjoy its richness and complexity more deeply. Perhaps your fear is neither good nor bad, just like a new spicy dish . . . or that odd music that's piqued your interest, which will surprise you much more if you listen to it without trying to define it or ascertain whether it's good or bad.

7. **Physical affection and storytelling with your loved ones will make you feel better**
   At the end of a long day of working or studying, we all look for ways to relax so we can fall asleep feeling good. Often

we turn to screens, movies, books, games, and all kinds of exercises to distract our minds from the remains of the day. It is worth trying to use that time to show your affection to those you love the most. They will feel protected and relaxed and, since affection is reflective, you will feel the same warm fuzzy feelings.

EPILOGUE

Feynman's mirror

Richard Feynman won the Nobel Prize in 1965 for pairing nineteenth-century physics with twentieth-century physics, electromagnetism with quantum mechanics. This conceptual leap, which consolidated him as one of the most extraordinary thinkers in science, was the result of simple diagrams that almost look as if they were drawn by a child. Feynman diagrams are still the best way to understand the interaction of particles in the quantum universe.

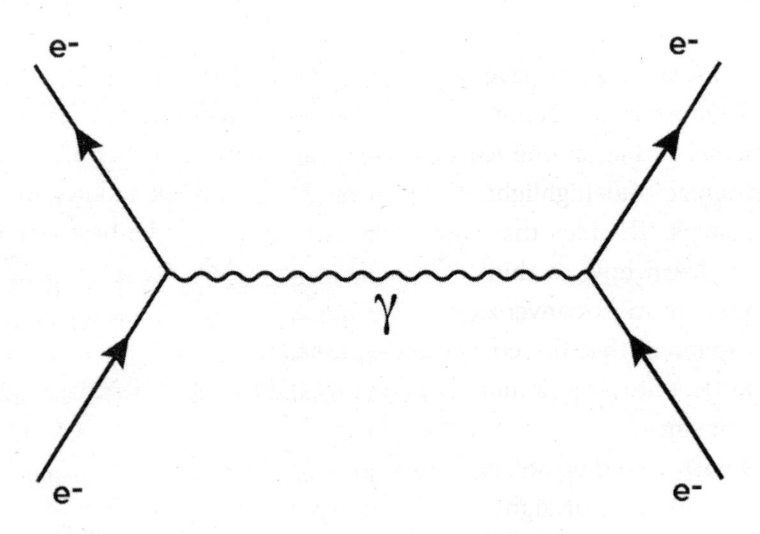

This was no accident. Feynman had an amazing virtue. He could understand any problem with simplicity and clarity, and express its every logical step in accessible language without an iota of ambiguity. Hence his diagrams. He conceived of a simple way to visually and mentally represent the most complex interactions of matter. Simple yet profound. In Feynman's world, those two concepts, instead of being opposites as is often believed, become almost redundant.

For this very reason, Feynman was an extraordinary teacher. Those who were lucky enough to study with him say it was the apex of their educational experience. He expressed his knowledge so clearly that they felt they were floating on air, the way the best music affects us. Feynman argued that this causality was reversed: he didn't teach well because he thought clearly. On the contrary: he thought clearly because he taught well. This was his formula for becoming a good scientist and, more generally, a good thinker: by teaching, in the clearest and simplest way possible, any problem that we have trouble understanding.

Feynman's method consists of a repeating loop: choose and define a research problem well. Think. Explain it to someone else, ideally a child. Find all the points in which the explanation falters, which is usually when we stammer or use big words to cover up what we don't know. Once we've identified those sticking points, we study them further and repeat the explanation until it flows perfectly. Only when that happens have we understood the problem.

Feynman's idea is not new. It is built on Seneca's maxim "*docendo discimus*," that is: one learns by teaching. Feynman's virtue was to recognize and highlight its importance. This book follows in his footsteps. The idea that conversing with others is the best way for us to learn how to think is present in every chapter. This doesn't happen in most conversations. It requires something Feynman took for granted: that the conversations not become a confrontation or a battle, but rather a mutual process of discovery. Talk to learn, not to convince.

I borrowed this tool, originally intended to explain facts of science and nature, and brought it to the discovery of our mental experience. We can emulate Feynman's process by explaining what we feel, or why we believe what we believe, or why we've made a decision. By explaining in simple words, as if we were speaking to a child, and paying attention to the weak points in our story. By re-examining ourselves to review the reasons that led us to feel an emotion or make a decision. By trying, with this introspective and narrative exercise, to reach deep into the profundity of our emotions.

In a way, I've done my own Feynman exercise here. In the process of sorting out these ideas, critically questioning when and why they're relevant, and how to put them into practice, I have found a better version of myself. A version I find more beautiful. I return to the beginning, in the invariably circular journey contained in every book: to the hope that you will have found these ideas useful, intriguing, and entertaining, and that they've stimulated you in the perpetual exercise of exploring the mercurial and captivating corners of our lives.

Acknowledgments

This book about conversations is the result of wonderful conversations; in this case the cobbler's children do wear shoes. It began to take shape in a course I gave at the Instituto Baikal on the power of words, in which some disjointed ideas were woven into a story. Soon after I started writing these ideas down, in conversation throughout the entire process with Marcos Trevisan, my companion on this voyage. We discussed, we wrote, we played, we lived together, and, above all, we laughed. Many of our jokes ended up in the footnotes. His narrative style, his talent for playing with words, and his friendship appear on every page.

Throughout the writing process I was constantly searching for ways to connect science with philosophy, history, and literature. This connection grew slowly over conversations with Mariana Noé in New York, with Christián Carman in Buenos Aires, and with Santiago Gerchunoff in Madrid. Mariana and I put together a conversation experiment joining science and philosophy that led to the last section of chapter 5, a practical investigation of emotions and their regulation that we wrote together. Christián was my guide on the journey to antiquity, with his contagious passion for the work of Saint Thomas and Aristotle. My conversation with Santiago were about more contemporary topics, but we discussed them in the ancient style, over long meals where philosophy, history, politics, soccer, and literature converged.

I conversed with Pedro Bekinschtein about memory, and with

Michael Posner, Philipp Kanske, and Adela Sáenz Cavia about emotional regulation. Melina Furman helped me to understand how memory and creativity are interwoven in the educational world. I exchanged ideas "à la Feynman" with Gerry Garbulsky and Emiliano Chamorro, in conversations where we shared an open enthusiasm for understanding and discovery.

Back in Madrid, I spent perfect days with Jacobo Bergareche, discussing his concept of the stations we return to on our life's timeline in a conversation that had the tone and intensity of our friendship. After four days in which we stopped only to eat and drink, I emerged exhausted and inspired by numerous stories that found their way into the pages of this book: about *nepenthes*, George Harrison, our first times, and about how damaging having only a single word to refer to different kinds of love can be. He and I wrote the cartoons that Javi Royo later captured, creating another conversation that blends humor, words, and drawings. It is amazing to meet someone for the first time and connect in dialogue, intent, and execution as if you've worked together for a lifetime. Borja Robert helped me to distill practical tools from what I'd written; these are featured at the end of each chapter. Isabel Garzo Ortega was instrumental in transforming the flow of arguments into diagrams that became a first version of the graphic summaries of each chapter. Milo and Noah filled my life with jokes that I transcribed in the footnotes, and always remind me how important it is to be able to laugh at yourself.

Finally, I thought I was done. But I still needed to do to the text what we do with our memory: edit it, rewrite it, delete it, correct it. Reconsolidate it. And that process began in my conversations with Iñigo Lomana, who gave me back a manuscript covered in red ink and suggestions. It continued with Santiago Llach, who asked me to replace the goalkeeper two minutes before the end of the game. He invoked Hemingway and graciously invited me to rewrite some of the passages to which I was most attached. It was about putting into practice what I'd written. Embracing difference, contradiction, different points of view. Gabriela Vigo revised the text very carefully and in her corrections she reminded me that my way of speaking and writing is an inevitable blend of my years lived in Spain and Argentina. It also helped me diagram and shape the book. Laura Angriman brought the illustrations to life. Anna Villada and I read the text through to make sure that, after so much talking, it still had the form of a book. You will be the final judge of that. And from there it made its way back to where it started: in the hands of Roberto Montes and Miguel Aguilar, my editors and friends on either side of the Atlantic, and to those conversations that launch an idea without even being able to imagine how it will evolve.

Much of the science I discuss in these pages I learned over my years of experiments in New York, Paris, and Buenos Aires. I would like to sincerely thank my colleagues in those factories of ideas: Charles Gilbert, Torsten Wiesel, Guillermo Cecchi, Marcelo Magnasco, Leopoldo Petreanu, Pablo Meyer, Eugenia Chiappe, Stanislas Dehaene, Jérôme Sackur, Laurent Cohen, Ghislaine Dehaene-Lambertz, Fabien Vinckier, Véronique Izard, Dan Ariely, Pablo Barttfeld, Ariel Zylberberg, Diego Fernandez Slezack, Facundo Carrillo, Joaquín Navajas, Cecilia Calero, Andrea Goldin, Juli Leone, Diego de la Hera, Diego Golombek, Agustín Ibañez, Rocco di Tella, Sidarta Ribeiro, Marcela Peña, Albert Costa, Mariano Sardon, Bruno Mesz, Gabriel Mindlin, Martin Beron de Astrada, Ramiro Freudenthal, Tristán Bekinschtein, Pablo Polosecki, Martin Elias Costa, Kathinka Evers, Carlos Diuk, Juan Frenkel, Andres Babino, Alejandro Maiche, Juan Valle Lisboa, Jacques Mehler, Marina Nespor, Antonio Battro, Sindey Strauss, Andrea Moro, Silvia Bunge, Susan Fitzpatrick, John Bruer, Elizabeth Spelke, Manuel Carreiras, Andrew Meltzoff, Andres Rieznik, Matías Lopez, Guillermo Solovey, Marie Amalric, Fede Zimmerman, Diego Shalom, Juan Kamienkowski, Adolfo Garcia, Hernan Makse, Alejo Salles, Santiago Figueira, Jacobo Sitt, Sergio Romano, Maria Luz Gadea, Julia Hermida, Edgar Altszyler, Andrea Slachevsky, Rafael di Tella, Ernesto Schargrodsky, Lionel Naccache, Liping Wang, Luis Martinez, Pierre Pica, Hal Pashler, Kim Shapiro, and John Duncan.

This project was born of my desire to find tools to improve some aspects of my life, primarily within my closest relationships. I offer my deepest gratitude to those loved ones, whom I don't need to mention here by name because they know who they are. My most loving gratitude, the most vivid and lovely gratitude I can possibly imagine. I am very lucky; I strive to remember that throughout my "*pequeña serenata diurna*," in an exercise of gratitude for what gives my life meaning. I am surrounded by good, kind, intelligent, fun, and endearing people, and I love them with my whole heart. This is a journey with them and for them. To paraphrase Montaigne: "They are the essence of this book."

Bibliography

Chapter 1

Asch, S. E. (1956). Studies of independence and conformity: I. A minority of one against a unanimous majority. *Psychological Monographs: General and Applied*, 70(9), 1–70.

Austin, J. L. (1962). *How to Do Things with Words.* Harvard University Press.

Borges, J. L. (1967–8). Norton Lectures: "A poet's creed." Harvard.

Borges, J. L. (1979). *Borges, Oral.* Belgra Editores/Editorial Belgrano.

Brady, W. J., Wills, J. A., Jost, J. T., Tucker, J. A., & Van Bavel, J. J. (2017). Emotion shapes the diffusion of moralized content in social networks. *Proceedings of the National Academy of Sciences*, 114(28), 7313–18.

Centola, D., Becker, J., Brackbill, D., & Baronchelli, A. (2018). Experimental evidence for tipping points in social convention. *Science*, 360(6393), 1116–19.

Claidière, N., Trouche, E., & Mercier, H. (2017). Argumentation and the diffusion of counter-intuitive beliefs. *Journal of Experimental Psychology: General*, 146(7), 1052.

Cohen, A. J. (2010). Music as a source of emotion in film. In P. N. Juslin & J. A. Sloboda (eds.), *Handbook of Music and Emotion: Theory, Research, Applications* (879–908). Oxford University Press.

Coleridge, S. T. (1984). *Biographia Literaria, or, Biographical sketches of my literary life and opinions.* Princeton University Press.

Davidai, S., & Gilovich, T. (2016). The headwinds/tailwinds asymmetry: an availability bias in assessments of barriers and blessings. *Journal of Personality and Social Psychology*, 111(6), 835.

Frederick, S. (2005). Cognitive reflection and decision making. *Journal of Economic Perspectives*, 19(4), 25–42.

Gombrich, E. (1968). *Meditations on a Hobby Horse*. Phaidon.

Heidegger, M. (1962). The Ontological Priority of the Question of Being. SCM.

Kahneman, D. (2011). *Thinking, Fast and Slow*. Penguin Books.

Keller, H. (1965). *The Story of My Life*. Dell.

Lee, K. (2016). "Can you really tell if a kid is lying?" TED.com.

Lumet, S. (1957). "12 Angry Men". Dubbed in Spanish and broadcast by RTVE.es, YouTube.com.

Maurette, P. (2021). *Por qué creemos en cuentos*. Clave Intelectual.

Merton, R. K. (1948). The self-fulfilling prophecy. *The Antioch Review*, 8(2), 193–210.

Mesz, B., Rodriguez Zivic, P. H., Cecchi, G. A., Sigman, M., & Trevisan, M. A. (2015). The music of morality and logic. *Frontiers in Psychology*, 6, 908.

Mesz, B., Trevisan, M. A., & Sigman, M. (2011). The taste of music. *Perception*, 40(2), 209–19.

Montaigne, M. (2004). *The Complete Essays*. Penguin UK.

Pearson, H. (2016). *The Life Project: The Extraordinary Story of our Ordinary Lives*. Catapult.

Pearson, H. (2017). "Lessons from the longest study on human development." TED.com.

Russell, B. (1980). Correspondence with Frege. In G. Frege *Philosophical and Mathematical Correspondence*. University of Chicago Press.

Schwarz, N., Bless, H., Strack, F., Klumpp, G., Rittenauer-Schatka, H., & Simons, A. (1991). Ease of retrieval as information: another look at the availability heuristic. *Journal of Personality and Social Psychology*, 61(2), 195.

Slater, M., Lotto, B., Arnold, M. M., & Sánchez-Vives, M. V. (2009). How we experience immersive virtual environments: the concept of presence and its measurement. *Anuario de Psicología*, 40(2), 193–210.

Soros, G. (2013). Fallibility, reflexivity, and the human uncertainty principle. *Journal of Economic Methodology*, 20:4, 309–29.

Todorov, T. (1987). *The Conquest of America: the problem of the other*. Harper & Row.

Tversky, A., & Kahneman, D. (1973). Availability: a heuristic for judging frequency and probability. *Cognitive Psychology*, 5(2), 207–32.

Vosoughi, S., Mohsenvand, M. N., & Roy, D. (2017). Rumor gauge: predicting the veracity of rumors on Twitter. *ACM Transactions on Knowledge Discovery from Data* (TKDD), 11(4), 1–36.

Vosoughi, S., Roy, D., & Aral, S. (2018). The spread of true and false news online. *Science*, 359(6380), 1146–51.

Vul, E., & Pashler, H. (2008). Measuring the crowd within: probabilistic representations within individuals. *Psychological Science*, 19(7), 645–7.

Chapter 2

Carey, S. (2000). The origin of concepts. *Journal of Cognition and Development*, 1(1), 37–41.

de la Hera, D. P., Sigman, M., & Calero, C. I. (2019). Social interaction and conceptual change pave the way away from children's misconceptions about the Earth. *npj Science of Learning*, 4(1), 1–12.

"El cerebro y Yo" (The Brain and I). Tiempo (2014). Via YouTube.com.

Epstein, D. (2013). *The Sports Gene: Inside the Science of Extraordinary Athletic Performance*. Penguin Books.

Galton, F. (1907). Vox populi (The wisdom of crowds). *Nature*, 75(7), 450–1.

Goldenberg, A., Cohen-Chen, S., Goyer, J. P., Dweck, C. S., Gross, J. J., & Halperin, E. (2018). Testing the impact and durability of a group malleability intervention in the context of the Israeli–Palestinian conflict. *Proceedings of the National Academy of Sciences*, 115(4), 696–701.

Haidt, J. (2001). The emotional dog and its rational tail: a social intuitionist approach to moral judgment. *Psychological Review*, 108(4), 814.

Halperin, E. (2008). Group-based hatred in intractable conflict in Israel. *Journal of Conflict Resolution*, 52(5), 713–36.

Mackay, C. (2012). *Extraordinary Popular Delusions and the Madness of Crowds*. Simon and Schuster.

Navajas, J., Heduan, F. Á., Garrido, J. M., Gonzalez, P. A., Garbulsky, G., Ariely, D., & Sigman, M. (2019). Reaching consensus in polarized moral debates. *Current Biology*, 29(23), 4124–9.

Navajas, J., Niella, T., Garbulsky, G., Bahrami, B., & Sigman, M. (2018). Aggregated knowledge from a small number of debates outperforms the wisdom of large crowds. *Nature Human Behaviour*, 2(2), 126–32.

Richter, C. P. (1957). 16. On the phenomenon of sudden death in animals and man. *Psychosomatic Medicine*, 19, 191–8.

Russell, J. (1991). Inventing the flat Earth. *History Today*, 41(8), 13–19.

Saramago, J. (1997). *Todos os nomes* (All the Names). Editora Companhia das Letras.

Vosniadou, S., & Brewer, W. F. (1992). Mental models of the earth: a study of conceptual change in childhood. *Cognitive Psychology*, 24(4), 535–85.

Yeager, D. S., & Dweck, C. S. (2012). Mindsets that promote resilience: when students believe that personal characteristics can be developed. *Educational Psychologist*, 47:4, 302–14.

Chapter 3

Akhtar, S., Justice, L. V., Morrison, C. M., & Conway, M. A. (2018). Fictional first memories. *Psychological Science*, 29(10), 1612–19.

Bachevalier, J. (1990). Ontogenetic development of habit and memory formation in primates. *Annals of the New York Academy of Sciences*, 608(1), 457–84.

Banerjee, A. V., & Duflo, E. (2007). The economic lives of the poor. *Journal of Economic Perspectives*, 21(1), 141–68.

Bender, C. L., Giachero, M., Comas-Mutis, R., Molina, V. A., & Calfa, G. D. (2018). Stress influences the dynamics of hippocampal structural remodeling associated with fear memory extinction. *Neurobiology of Learning and Memory*, 155, 412–21.

Bentham, J. (1780). *Le panoptique.*

Benzenhöfer, U., & Passie, T. (2010). Rediscovering MDMA (ecstasy): the role of the American chemist Alexander T. Shulgin. *Addiction*, 105(8), 1355–61.

Bergareche, J. (2019). *Estaciones de regreso*. Círculo de Tiza.

Brown, R., & McNeill, D. (1966). The "tip of the tongue" phenomenon. *Journal of Verbal Learning and Verbal Behavior*, 5(4), 325–37.

Carrier, M., & Pashler, H. (1992). The influence of retrieval on retention. *Memory & Cognition*, 20(6), 633–42.

Cepeda, N. J., Vul, E., Rohrer, D., Wixted, J. T., & Pashler, H. (2008). Spacing effects in learning: A temporal ridgeline of optimal retention. *Psychological Science*, 19(11), 1095–1102.

Chernev, A., Böckenholt, U., & Goodman, J. (2015). Choice overload: a conceptual review and meta-analysis. *Journal of Consumer Psychology*, 25(2), 333–58.

Chialvo, D. R. (2003). How we hear what is not there: a neural mechanism for the missing fundamental illusion. *Chaos: An Interdisciplinary Journal of Nonlinear Science*, 13(4), 1226–30.

Courage, M. L., Edison, S. C., & Howe, M. L. (2004). Variability in the early development of visual self-recognition. *Infant Behavior and Development*, 27(4), 509–32.

Deleuze, G., Guattari, F. (1991). *¿Qué es la filosofía?* (What is Philosophy?). Anagrama.

Deleuze, G. (2016). *Pintura: el concepto de diagrama. Clases en versión castellana del curso dictado por Deleuze en la Universidad de Vincennes en 1981*, ed. Cactus.

Diuk, C., Fernandez Slezak, D., Raskovsky, I., Sigman, M., & Cecchi, G. A. (2012). A quantitative philology of introspection. *Frontiers in Integrative Neuroscience*, 6, 80.

Dresler, M., Shirer, W. R., Konrad, B. N., Müller, N. C., Wagner, I. C., Fernández, G., . . . & Greicius, M. D. (2017). Mnemonic training reshapes brain networks to support superior memory. *Neuron*, 93(5), 1227–35.

Foer, J. (2011). *Moonwalking with Einstein*. Penguin Books.

Foucault, M. (1975). *Surveiller et Punir: Naissance de la prison*. Editions Gallimard.

Freud, S. (1905/1963). Introductory lectures on psychoanalysis [translated]. In J. Strachey (ed.), *The Standard Edition of the Complete Psychological Works of Sigmund Freud* (Vol. 15, pp. 199–201). Hogarth Press.

Freudenmann, R. W., Öxler, F., & Bernschneider-Reif, S. (2006). The origin of MDMA (ecstasy) revisited: the true story reconstructed from the original documents. *Addiction*, 101(9), 1241–5.

Furman, M. (2019). *Enseñar distinto, guía para innovar sin perderse en el camino*. Siglo Veintiuno Editores.

Gallup Jr., G. G., Anderson, J. R., & Shillito, D. J. (2002). In M. Bekoff, C. Allen, & G. M. Burghardt (eds.), *The Cognitive Animal: Empirical and theoretical perspectives on animal cognition* (325–33). MIT Press.

Hafting, T., Fyhn, M., Molden, S., Moser, M. B., & Moser, E. I. (2005). Microstructure of a spatial map in the entorhinal cortex. *Nature*, 436(7052), 801–6.

Hall, W., & Carter, A. (2007). Debunking alarmist objections to the pharmacological prevention of PTSD. *The American Journal of Bioethics*, 7(9), 23–5.

Harley, T. A., & Bown, H. E. (1998). What causes a tip-of-the-tongue state? Evidence for lexical neighbourhood effects in speech production. *British Journal of Psychology*, 89(1), 151–74.

"How Paul McCartney wrote 'Yesterday.'" The Beatles Anthology, Episode Four. Apple Corps, 2003. YouTube.com.

Howe, M. L., Courage, M. L., & Edison, S. C. (2003). When autobiographical memory begins. *Developmental Review*, 23(4), 471–94.

Howe, M. L., Threadgold, E., Wilkinson, S., Garner, S. R., & Ball, L. J. (2017). The positive side of memory illusions: new findings about how false memories facilitate reasoning and problem solving. In R. A. Nash & J. Ost, (eds.), *False and Distorted Memories* (130–42). Routledge/Taylor & Francis Group.

Izard, V., Sann, C., Spelke, E. S., & Streri, A. (2009). Newborn infants perceive abstract numbers. *Proceedings of the National Academy of Sciences*, 106(25), 10382–5.

Jaynes, J. (1976). *The Origins of Consciousness in the Breakdown of the Bicameral Mind*. Mariner Books.

Jaynes, J. (1986). Consciousness and the voices of the mind. *Canadian Psychology/Psychologie Canadienne*, 27(2), 128.

Josselyn, S. A., & Tonegawa, S. (2020). Memory engrams: recalling the past and imagining the future. *Science*, 367(6473).

Kellogg, W. N., & Kellogg, L. A. (1933). *The Ape and the Child: A Study of Environmental Influence Upon Early Behavior*. Whittlesey House.

Legge, E. L., Madan, C. R., Ng, E. T., & Caplan, J. B. (2012). Building a memory palace in minutes: Equivalent memory performance using virtual versus conventional environments with the Method of Loci. *Acta Psychologica*, 141(3), 380–90.

Liu, X., Ramirez, S., Pang, P. T., Puryear, C. B., Govindarajan, A., Deisseroth, K., & Tonegawa, S. (2012). Optogenetic stimulation of a hippocampal engram activates fear memory recall. *Nature*, 484(7394), 381–5.

Loftus, E. F., & Palmer, J. C. (1996). Eyewitness testimony. In P. Banyard & A. Grayson, (eds.), *Introducing Psychological Research* (305–9). Palgrave.

Lopez-Rosenfeld, M., Calero, C. I., Fernandez Slezak, D., Garbulsky, G., Bergman, M., Trevisan, M., & Sigman, M. (2015). Neglect in human communication: quantifying the cost of cell-phone interruptions in face to face dialogs. *PLOS One*, 10(6), e0125772.

Marton, F., & Säljö, R. (1976). On qualitative differences in learning: I. Outcome and process. *British Journal of Educational Psychology*, 46(1): 4–11.

McGaugh, J. L. (2000). Memory: a century of consolidation. *Science*, 287(5451), 248–51.

Mednick, S. (1962). The associative basis of the creative process. *Psychological Review*, 69(3), 220.

Meltzoff, A. N. (1995). What infant memory tells us about infantile amnesia: long-term recall and deferred imitation. *Journal of Experimental Child Psychology*, 59(3), 497–515.

Misanin, J. R., Miller, R. R., & Lewis, D. J. (1968). Retrograde amnesia produced by electroconvulsive shock after reactivation of a consolidated memory trace. *Science*, 160(3827), 554–5.

Mithoefer, M. C., Wagner, M. T., Mithoefer, A. T., Jerome, L., & Doblin, R. (2011). The safety and efficacy of ±3,4-methylenedioxymethamphetamine-assisted psychotherapy in subjects with chronic, treatment-resistant posttraumatic stress disorder: the first randomized controlled pilot study. *Journal of Psychopharmacology*, 25(4), 439–52.

Morrison, C. M., & Conway, M. A. (2010). First words and first memories. *Cognition*, 116(1), 23–32.

Offit, P. A. (2007). *Vaccinated: One Man's Quest to Defeat the World's Deadliest Diseases*. Smithsonian Books/HarperCollins.

Perner, J., & Ruffman, T. (1995). Episodic memory and autonoetic consciousness: developmental evidence and a theory of childhood amnesia. *Journal of Experimental Child Psychology*, 59(3), 516–48.

Polgár, J. (2016). "Giving checkmate is always fun." In "The dreams that define us: The global talks in Session 10 of TED2016." Blog.TED.com.

Reutskaja, E., Lindner, A., Nagel, R., Andersen, R. A., & Camerer, C. F. (2018). Choice overload reduces neural signatures of choice set value in dorsal striatum and anterior cingulate cortex. *Nature Human Behaviour*, 2(12), 925–35.

Roediger, H. L., & McDermott, K. B. (1995). Creating false memories: remembering words not presented in lists. *Journal of Experimental Psychology: Learning, Memory, and Cognition*, 21(4), 803.

Roy, B. C., Frank, M. C., DeCamp, P., Miller, M., & Roy, D. (2015). Predicting the birth of a spoken word. *Proceedings of the National Academy of Sciences*, 112(41), 12663–8.

Ryan, T. J., Roy, D. S., Pignatelli, M., Arons, A., & Tonegawa, S. (2015). Engram cells retain memory under retrograde amnesia. *Science*, 348(6238), 1007–13.

Schafe, G. E., & LeDoux, J. E. (2000). Memory consolidation of auditory Pavlovian fear conditioning requires protein synthesis and protein kinase A in the amygdala. *Journal of Neuroscience*, 20(18), RC96.

Schwartz, B. (2004). *The Paradox of Choice: Why More is Less*. Harper Perennial.

Semon, R. W. (1921). *The Mneme*. George Allen & Unwin Ltd.; Macmillan.

Shofner, W. P. (2005). Comparative aspects of pitch perception. In C. J. Plack, R. R. Fay, A. J. Oxenham, & A. N. Popper (eds.), *Pitch* (pp. 56–98). Springer.

Simner, J., & Ward, J. (2006). The taste of words on the tip of the tongue. *Nature*, 444(7118), 438.

Spanos, N. P., Burgess, C. A., & Burgess, M. F. (1994). Past-life identities, UFO abductions, and satanic ritual abuse: the social construction of memories. *International Journal of Clinical and Experimental Hypnosis*, 42(4), 433–46.

Squire, L. R. (2004). Memory systems of the brain: a brief history and current perspective. *Neurobiology of Learning and Memory*, 82(3), 171–7.

Stickgold, R. (2005). Sleep-dependent memory consolidation. *Nature*, 437(7063), 1272–8.

Tonegawa, S., Morrissey, M. D., & Kitamura, T. (2018). The role of engram cells in the systems consolidation of memory. *Nature Reviews Neuroscience*, 19(8), 485–98.

Treisman, A., & Schmidt, H. (1982). Illusory conjunctions in the perception of objects. *Cognitive Psychology*, 14(1), 107–41.

Tulving, E. (1993). What is episodic memory? *Current Directions in Psychological Science*, 2(3), 67–70.

Wagon, S. (1993). *The Banach–Tarski Paradox* (No. 24). Cambridge University Press.

Wilson, A., & Ross, M. (2003). The identity function of autobiographical memory: time is on our side. *Memory*, 11(2), 137–49.

Yates, F. A. (1992). *The Art of Memory* (Vol. 64). Random House UK.

Young, C., & Butcher, R. (2020). Propranolol for post-traumatic stress disorder: a review of clinical effectiveness. Canadian Agency for Drugs and Technologies in Health: CADTH Rapid Response Reports.

Zermelo, E. (1908). "Untersuchungen über die Grundlagen der Mengenlehre I" (Investigations into the Foundations of Set Theory I). *Mathematische Annalen* 65: 261–81.

Chapter 4

Bar, M., & Biederman, I. (1998). Subliminal visual priming. *Psychological Science*, 9(6), 464–8.

Barrett, L. F. (2012). Emotions are real. *Emotion*, 12(3), 413.

Barrett, L. F. (2017). *How Emotions Are Made: The Secret Life of the Brain*. Houghton Mifflin Harcourt.

Berra, T. M. (2013). *Darwin and His Children: His Other Legacy*. Oxford University Press.

Borges, J. L. (1952). El idioma analítico de John Wilkins. *Otras inquisiciones*, 2.

Chomsky, N. (1964). [The Development of Grammar in Child Language]: Discussion. *Monographs of the Society for Research in Child Development*, 35–42.

Di Tella, R., et al. (2019). "Pecados capitales." El Gato y la Caja.

Ekman, P. (1992). An argument for basic emotions. *Cognition & Emotion*, 6(3–4), 169–200.

Everett, D. (2017). *How Language Began: The Story of Humanity's Greatest Invention*. Profile Books.

Goodenough, U. W. (1991). Deception by pathogens. *American Scientist*, 79(4), 344–55.

Henrich, J., Heine, S. J., & Norenzayan, A. (2010). The weirdest people in the world? *Behavioral and Brain Sciences*, 33(2–3), 61–83.

Hess, U., & Thibault, P. (2009). Darwin and emotion expression. *American Psychologist*, 64(2), 120.

Horikawa, T., Tamaki, M., Miyawaki, Y., & Kamitani, Y. (2013). Neural decoding of visual imagery during sleep. *Science*, 340(6132), 639–42.

Koenig, J. (2021). *The Dictionary of Obscure Sorrows*. Simon and Schuster.

Kuhl, P. K. (2004). Early language acquisition: cracking the speech code. *Nature Reviews Neuroscience*, 5(11), 831–43.

Kuhl, P. K., Williams, K. A., Lacerda, F., Stevens, K. N., & Lindblom, B. (1992). Linguistic experience alters phonetic perception in infants by 6 months of age. *Science*, 255(5044), 606–8.

Lakoff, G., & Johnson, M. (2008). *Metaphors We Live By*. University of Chicago Press.

Landauer, T. K., & Dumais, S. T. (1997). A solution to Plato's problem: the latent semantic analysis theory of acquisition, induction, and representation of knowledge. *Psychological Review*, 104(2), 211.

Leone, M. J., Salles, A., Pulver, A., Golombek, D. A., & Sigman, M. (2018). Time drawings: spatial representation of temporal concepts. *Consciousness and Cognition*, 59, 10–25.

Lieberman, P. (2007). The evolution of human speech: its anatomical and neural bases. *Current Anthropology*, 48(1), 39–66.

Núñez, R. E., & Sweetser, E. (2006). With the future behind them: convergent evidence from Aymara language and gesture in the crosslinguistic comparison of spatial construals of time. *Cognitive Science*, 30(3), 401–50.

Peirce, C. S. (1991). *Peirce on Signs: Writings on Semiotic*. UNC Press Books.

Plutchik, R. (1991). *The Emotions*. University Press of America.

Plutchik, R. (2001). The nature of emotions: human emotions have deep evolutionary roots, a fact that may explain their complexity and provide tools for clinical practice. *American Scientist*, 89(4), 344–350.

Quine, W. V. O. (2013). *Word and Object*. MIT Press.

Ramachandran, V. S., & Hubbard, E. M. (2001). Synaesthesia: a window into perception, thought and language. *Journal of Consciousness Studies*, 8(12), 3–34.

Rosaldo, R. (2004). Grief and a Headhunter's Rage. In A. C. G. M. Robben (ed.), *Death, Mourning, and Burial: A Cross-Cultural Reader*, 167–78.

Russell, J. A., & Mehrabian, A. (1977). Evidence for a three-factor theory of emotions. *Journal of Research in Personality*, 11(3),273–94.

Scott, J. P. (1958). *Animal Behavior*. University of Chicago Press.

Wittgenstein, L. (1921). *Tractatus logico-philosophicus*. Routledge.

Wnuk, E., & Majid, A. (2014). Revisiting the limits of language: the odor lexicon of Maniq. *Cognition*, 131(1), 125–38.

Yang, Z., & Tong, E. M. (2010). The effects of subliminal anger and sadness primes on agency appraisals. *Emotion*, 10(6), 915.

Chapter 5

Adolph, K. E. (2000). Specificity of learning: why infants fall over a veritable cliff. *Psychological Science*, 11(4), 290–5.

Anderson, R. A. (2006). Influenza vaccine antibody response and loneliness. *Townsend Letter: The Examiner of Alternative Medicine*, (281), 143.

Barss, P. (1984). Injuries due to falling coconuts. *The Journal of Trauma*, 24(11), 990–1.

Baumeister, R. F., DeWall, C. N., Ciarocco, N. J., & Twenge, J. M. (2005). Social exclusion impairs self-regulation. *Journal of Personality and Social Psychology*, 88(4), 589.

Cacioppo, J. T., Ernst, J. M., Burleson, M. H., McClintock, M. K., Malarkey, W. B., Hawkley, L. C., . . . & Berntson, G. G. (2000). Lonely traits and concomitant physiological processes: the MacArthur social neuroscience studies. *International Journal of Psychophysiology*, 35(2–3), 143–54.

Carroll, J. L. (1990). The relationship between humor appreciation and perceived physical health. *Psychology: A Journal of Human Behaviour*, 27, 34–7.

Chemaly, S. (2019). "The power of women's anger." TED.com.

Cole, S. W., Hawkley, L. C., Arevalo, J. M., Sung, C. Y., Rose, R. M., & Cacioppo, J. T. (2007). Social regulation of gene expression in human leukocytes. *Genome Biology*, 8(9), 1–13.

Dostoevsky, F. (1863). *Apuntes de invierno sobre impresiones de verano* (Winter Notes on Summer Impressions). "Intente no pensar en un oso polar y verá a ese maldito animal a cada minute" (Try not to think of a polar bear and you will see the damned animal every minute).

Erzen, E., & Çikrikci, Ö. (2018). The effect of loneliness on depression: a meta-analysis. *International Journal of Social Psychiatry*, 64(5), 427–35.

Etkin, A., Egner, T., & Kalisch, R. (2011). Emotional processing in anterior cingulate and medial prefrontal cortex. *Trends in Cognitive Sciences*, 15(2), 85–93.

Gallagher, M., & Chiba, A. A. (1996). The amygdala and emotion. *Current Opinion in Neurobiology*, 6(2), 221–7.

Hawkley, L. C., Thisted, R. A., Masi, C. M., & Cacioppo, J. T. (2010). Loneliness predicts increased blood pressure: five-year cross-lagged analyses in middle-aged and older adults. *Psychology and Aging*, 25(1), 132.

Hommel, B., Pratt, J., Colzato, L., & Godijn, R. (2001). Symbolic control of visual attention. *Psychological Science*, 12(5), 360–5.

James, W. (1863). *The Principles of Psychology.*

John, O. P., & Gross, J. J. (2004). Healthy and unhealthy emotion regulation: personality processes, individual differences, and life span development. *Journal of Personality*, 72(6), 1301–34.

Judith, G. R. (1995). Long-term survival with AIDS and the role of community. *AIDS, Identity, and Community*, 2, 169.

Kanai, R., Bahrami, B., Duchaine, B., Janik, A., Banissy, M. J., & Rees, G. (2012). Brain structure links loneliness to social perception. *Current Biology*, 22(20), 1975–9.

Klapp, S. T. (2015). One version of direct response priming requires automatization of the relevant associations but not awareness of the prime. *Consciousness and Cognition*, 34, 163–75.

Kraft, T. L., & Pressman, S. D. (2012). Grin and bear it: the influence of manipulated facial expression on the stress response. *Psychological Science*, 23(11), 1372–8.

McEwen, B. S., & Seeman, T. (1999). Protective and damaging effects of mediators of stress: elaborating and testing the concepts of allostasis and allostatic load. *Annals of the New York Academy of Sciences*, 896(1), 30–47.

Méndez-Bértolo, C., Moratti, S., Toledano, R., Lopez-Sosa, F., Martínez-Alvarez, R., Mah, Y. H., . . . & Strange, B. A. (2016). A fast pathway for fear in human amygdala. *Nature Neuroscience*, 19(8), 1041–9.

Noah, T., Schul, Y., & Mayo, R. (2018). When both the original study and its failed replication are correct: feeling observed eliminates the facial-feedback effect. *Journal of Personality and Social Psychology*, 114(5), 657.

Ochsner, K. N., & Gross, J. J. (2005). The cognitive control of emotion. *Trends in Cognitive Sciences*, 9(5), 242–9.

Ochsner, K. N., Bunge, S. A., Gross, J. J., & Gabrieli, J. D. (2002). Rethinking feelings: an FMRI study of the cognitive regulation of emotion. *Journal of Cognitive Neuroscience*, 14(8), 1215–29.

Posner, M. I., & Petersen, S. E. (1990). The attention system of the human brain. *Annual Review of Neuroscience*, 13(1), 25–42.

Reuss, H., Kiesel, A., Kunde, W., & Wühr, P. (2012). A cue from the unconscious – masked symbols prompt spatial anticipation. *Frontiers in Psychology*, 3, 397.

Richards, J. M., & Gross, J. J. (1999). Composure at any cost? The cognitive consequences of emotion suppression. *Personality and Social Psychology Bulletin*, 25(8), 1033–44.

Rosenberg, E. L., Ekman, P., Jiang, W., Babyak, M., Coleman, R. E., Hanson, M., . . . & Blumenthal, J. A. (2001). Linkages between facial expressions of anger and transient myocardial ischemia in men with coronary artery disease. *Emotion*, 1(2), 107.

Strack, F., Martin, L. L., & Stepper, S. (1988). Inhibiting and facilitating conditions of the human smile: a nonobtrusive test of the facial feedback hypothesis. *Journal of Personality and Social Psychology*, 54(5), 768.

Sutin, A. R., Stephan, Y., Luchetti, M., & Terracciano, A. (2020). Loneliness and risk of dementia. *The Journals of Gerontology: Series B*, 75(7), 1414–22.

Uchino, B. N., Cacioppo, J. T., & Kiecolt-Glaser, J. K. (1996). The relationship between social support and physiological processes: a review with emphasis on underlying mechanisms and implications for health. *Psychological Bulletin*, 119, 488–531.

Vasarhelyi, E. C., & Chin, J. (2018). *Free Solo*. National Geographic Documentary Films.

Vigneau, M., Beaucousin, V., Hervé, P. Y., Duffau, H., Crivello, F., Houde, O., . . . & Tzourio-Mazoyer, N. (2006). Meta-analyzing left hemisphere language areas: phonology, semantics, and sentence processing. *Neuroimage*, 30(4), 1414–32.

Völlm, B. A., Taylor, A. N., Richardson, P., Corcoran, R., Stirling, J., McKie, S., . . . & Elliott, R. (2006). Neuronal correlates of theory of mind and empathy: a functional magnetic resonance imaging study in a nonverbal task. *Neuroimage*, 29(1), 90–8.

Wagenmakers, E. J., Beek, T., Dijkhoff, L., Gronau, Q. F., Acosta, A., Adams Jr., R. B., . . . & Zwaan, R. A. (2016). Registered Replication Report: Strack, Martin, & Stepper (1988). *Perspectives on Psychological Science*, 11(6), 917–28.

Chapter 6

Angeles, L. (2010). Children and life satisfaction. *Journal of Happiness Studies*, 11(4), 523–38.

Barttfeld, P., Wicker, B., McAleer, P., Belin, P., Cojan, Y., Graziano, M., . . . & Sigman, M. (2013). Distinct patterns of functional brain connectivity correlate with objective performance and subjective beliefs. *Proceedings of the National Academy of Sciences*, 110(28), 11577–82.

Breines, J. G., Thoma, M. V., Gianferante, D., Hanlin, L., Chen, X., & Rohleder, N. (2014). Self-compassion as a predictor of interleukin-6 response to acute psychosocial stress. *Brain, Behavior, and Immunity*, 37, 109–14.

Cicero. *De finibus, I*, 30, on Epicurus.

Frith, C. D. (2014). *The Cognitive Neuropsychology of Schizophrenia*. Psychology Press.

Gottman, J. M., & Levenson, R. W. (2000). The timing of divorce: predicting when a couple will divorce over a 14-year period. *Journal of Marriage and Family*, 62(3), 737–45.

Killingsworth, M. A., & Gilbert, D. T. (2010). A wandering mind is an unhappy mind. *Science*, 330(6006), 932.

Kober, H., Buhle, J., Weber, J., Ochsner, K. N., & Wager, T. D. (2019). Let it be: mindful acceptance down-regulates pain and negative emotion. *Social Cognitive and Affective Neuroscience*, 14(11), 1147–58.

Leary, M. R., Tate, E. B., Adams, C. E., Batts Allen, A., & Hancock, J. (2007). Self-compassion and reactions to unpleasant self-relevant events: the implications of treating oneself kindly. *Journal of Personality and Social Psychology*, 92(5), 887.

Levenson, R. W., Carstensen, L. L., & Gottman, J. M. (1993). Long-term marriage: age, gender, and satisfaction. *Psychology and Aging*, 8(2), 301.

Mason, M. F., Norton, M. I., Van Horn, J. D., Wegner, D. M., Grafton, S. T., & Macrae, C. N. (2007). Wandering minds: the default network and stimulus-independent thought. *Science*, 315(5810), 393–5.

Neff, K. D. (2003). The development and validation of a scale to measure self-compassion. *Self and Identity*, 2(3), 223–50.

Neff, K. D., Long, P., Knox, M. C., Davidson, O., Kuchar, A., Costigan, A., . . . & Breines, J. G. (2018). The forest and the trees: examining the association of self-compassion and its positive and negative components with psychological functioning. *Self and Identity*, 17(6), 627–45.

Norton, M. I., Mochon, D., & Ariely, D. (2012). The IKEA effect: when labor leads to love. *Journal of Consumer Psychology*, 22(3), 453–60.

Nozick, R. (2013). The experience machine. In R. Shafer-Landau (ed.), *Ethical Theory: An Anthology* 264–5.

Patoilo, M. S., Berman, M. E., & Coccaro, E. F. (2021). Emotion attribution in intermittent explosive disorder. *Comprehensive Psychiatry*, 106, 152229.

Sheridan, M. A., Fox, N. A., Zeanah, C. H., McLaughlin, K. A., & Nelson, C. A. (2012). Variation in neural development as a result of exposure to institutionalization early in childhood. *Proceedings of the National Academy of Sciences*, 109(32), 12927–32.

White, L. K., Booth, A., & Edwards, J. N. (1986). Children and marital happiness: why the negative correlation? *Journal of Family Issues*, 7(2), 131–47.

Index

Adolph, Karen 199–200
AIDS pandemic, loneliness in 177–8, 244
air accidents 52
Altom, Jason 233–4
amnesia, infantile 77, 102–6
"Analytical Language of John Wilkins, The" (Borges essay) 142–5
anger
  of the Hulk 181, 204
  at ourselves or our children 188–9, 227–8, 229, 235
  purpose and resignification of 204, 208, 218–19
  relationship to fear 150, 167, 181, 229
  relationship to sadness 152, 204, 217
  wrath (deadly sin of) 153–4, 197
anisomycin (protein-blocking drug) 111–12
Aquinas, Saint Thomas 145, 149–50, 167, 181
Ariely, Dan 231–2
Aristotle 145, 149, 194, 205, 234
aromas, words describing 139–41
"Arrow of Time, The" (Goldberg photographic series) 98
artificial intelligence 132, 157, 158–9
Asch, Solomon 28–9, 33
attentional system 179–80, 195–6, 209–11. *See also* distraction
Austin, John 19
autobiographical stories. *See* life story formation
automatic *versus* logical thinking 11, 20–3, 26, 36
Aventura 19
axiom of choice 87–8
Aymara people 134–5, 137
Aztec people 135, 152

bad conversations. *See also* good conversations
  intimidation cause 33, 51–2

bad conversations (*contd.*)
mobile phone distraction cause 183
negative self-talk 4, 20, 65–6, 188, 227, 230–1, 241
on social media 33, 35–6, 48–9
social pressure cause 28–9, 33, 41–2
Bahrami, Bahador 177
Bannister, Roger 67
Baricco, Alessandro 182, 183
Baronchelli, Andrea 31, 32, 33
Barrett, Lisa Feldman 147, 152
Barss, Peter 201
Barttfeld, Pablo 252
Bay of Pigs invasion (1961) 41–2, 50
Bekinschtein, Pedro 116, 117
Bell, Alexander Graham 278
Bergareche, Jacobo 85, 105
bias. *See* fallibility
Biles, Simone 234
Bioy Casares, Adolfo 132
Boétie, Étienne de la 176
Borges, Jorge Luis 15, 132, 142–5
Bosch, Hieronymus 153
Boston Marathon bombing (2013) 12
Bown, Helen 96, 98
Brady, William 13
breathing 251–2
Brown, Roger 96
Bugkalot people 162–3, 237
Buridan, Jean 87
Burma, Nestor (fictional detective) 123–4
C.K., Louis 102
Cacioppo, John 179–80, 248
Calero, Cecilia 59–61
Calliope 80
Carey, Susan 107–8
Carlin, John 239–40
Carman, Christián 163, 166
Castro, Fidel 41–2
Chapman, Graham 238
Chemaly, Soraya 204
chess players 47, 99
Chesterton, G. K. 142
children
anger directed at 189, 227–8, 229–30, 235
attentional system of infants 196
Darwin's studies of 146
emotional neglect, effects on 248–9
fear, acquisition of 199–200
flat Earth belief of 58–61
games and fictions of 14, 15–16, 86, 118, 219
with growth or inflexible mentalities 65–6
infantile amnesia and first memories 77, 102–8
language acquisition by 98–101, 132, 138–9, 156–9
Chomsky, Noam 156
Cicero 34
cinema 14, 212–13
Cleese, John 238
cocktail party effect 179–80
cognitive bias. *See* fallibility
Cohen, Leonard 214
color perceptions 109–10, 118, 141
Columbus, Christopher 58

compassion. *See also* good conversations; love
absence of (loneliness) 175, 177–80, 221, 244, 248–9
absence of (negative self-talk) 4, 20, 65–6, 188, 227, 230–1, 241
advice on 257–9
changing perspective for 230–8, 247, 258
emotional governance by 184, 188–9, 227–8, 229–30, 245
expressed as attention or affection 132, 178, 221, 247–50, 258–9
humor for 230, 238–40
meditation exercise for 251–6
self-compassion 240–7, 255–6
computer files 106, 113, 114–15
"confident grays"/"Montaignes" (consensus makers) 56–7, 63–4, 71
conflict resolution
consensus makers needed for 56–7, 63–4, 71
growth mentality needed for 64–5, 71
consolidation/reconsolidation processes of memory 111–14, 116–17
contagion effect of emotions 48, 165–6, 190–4, 222, 230
conversation, power of. *See* bad conversations; good conversations
Corey, Elias 233–4
Cortázar, Julio 201, 202
cortisol 116, 117, 118
creativity and memory connection. *See also* life story formation
ancient Greek view 80–2, 91
creative techniques for memorizing 91–4, 125–6
false memories, creativity of 78, 120–4, 165
formal education, role in 79, 88–91
paradox of choice 86–8, 125
remembering words 95–8
triggers for 83–6, 91, 125
writing, effect on 82–3, 91
crowd experiments. *See* group decision-making experiments
Cuban missile crisis (1962) 42–3

Dante Alighieri 132, 153–4
Darwin, Charles 145–6
death, fear of 237–40
decision-making. *See also* reasoning
excessive choice, effect on 86–8, 125
group experiments in *See* group decision-making experiments
intuitive 43–4
minority views, potentially persuasive force 29–32, 61
deep knowledge/learning 89–90, 94–5, 125
Dehaene, Stanislas 241
Deleuze, Gilles 86
Descartes, René 143, 145, 149
dialogue. *See* bad conversations; good conversations

digital files 106, 113, 114–15
disgust 215–16
distraction
attentional system 179–80, 195–6, 209–11
emotional governance by 184, 187, 194–8, 221
social media causing 183, 187, 197
stress caused by 176, 197–8, 206–7, 222
Divine Comedy, The (Dante) 153–4
Dolina, Alejandro 79
dreams 80–2, 83, 88, 164
Drexler, Jorge 63, 238, 242
Dumais, Susan 156, 157–9
Dweck, Carol 65–6

Edict of Nantes (1598) 57
editing memories 77–9, 95, 113–14, 116–17, 126
Einstein, Albert 19, 133
Ekman, Paul 146–7
El cerebro y yo (television program) 69
El Erudito (board game) 44–5
El Guerrouj, Hicham 67
emotions. *See also* particular emotions
as bodily and cerebral activity 163–8, 170, 189–92
contagion effect of 48, 165–6, 190–4, 222, 230
fake news dissemination and 13
logical thinking affected by 31, 36, 208
Plutchik's wheel 151–2, 153, 159–60, 203
role in remembering 115–18
seven deadly sins 153–4, 197
taxonomies of 149–51
emotions, governing
advice on 221–2
by compassion *See* compassion
by distraction *See* distraction
with good conversations 175, 177–80, 221
heroes of emotional regulation 184–6
by induction 184, 186–7, 189–94, 222
by resignification *See* emotions, resignification (reappraisal) of
by suppression 176, 206–7, 209–11, 222
ungoverned emotions in fiction 181–2
emotions, resignification (reappraisal) of
advice on 222
brain activity involved in 206–13
by changing words 170, 176, 184
by exploring emotional middle grounds 159–60, 170, 204–6
fear recast as excitement example 132, 187–8, 198–200
individual and cultural factors affecting 200–4
resignifying particular emotions 214–20

emotions, words describing
complex emotions, finding words for 159–63, 169–70, 205–6
fallibility of 18–19, 147
precision and nuance lacking 11, 141–2, 161–2
reflexive potential 17, 19, 35, 131, 148–9, 168, 203
resignification of *See* emotions, resignification (reappraisal) of
universal expressions, searches for 145–7
empathy 242, 254. *See also* compassion
engrams 113, 114, 116, 208
Epicurus 236–7
episodic memory 105, 106
Epstein, David 67–8
Eratosthenes 58
explicit memory 105
eyewitness testimony 110–11, 120

fake news
about ourselves 11, 16–20, 77–8
on social media 12–13
fallibility. *See also* reflexivity (self-fulfilling prophecies)
automatic thinking cause 11, 20–3, 26, 36
of emotion statements 18–19, 147
false memories 78, 108–11, 119–24, 126, 165
lack of evidence cause 24–6, 109–10, 157
overestimating successes, failures, and problems 188–9, 227, 231–4, 246
recognition as self-help strategy 33, 35
reflexivity, interaction with 17–19, 26
false memories 78, 108–11, 119–24, 126, 165
Fast, Howard 62
fear
anger, relationship to 150, 167, 181, 229
of death 237–9
inhibiting effect of 33, 51–2, 86
physical signs 164, 167–8, 214
purpose and resignification of 132, 187–8, 198–200, 215, 218
versus reason 68–9, 200–2
Federer, Roger 185, 186
Feferovich, Sergio 163
Feldman Barrett, Lisa 147, 152
Feynman, Richard 260–2
fictional worlds
amphibious (chosen) existence in 13–15, 110, 232–3, 235, 257
jealousy creating 182, 220
memory, role in creating *See* creativity and memory connection
reality intruding upon 15–16, 17–18
financial market dynamics 17–18, 148
flat Earth belief 58–61

Fonda, Henry (in *Twelve Angry Men*) 29–31, 32, 33, 61
forgetting and remembering 95–8, 106, 114–18, 139
Fox, Nathan 248
Frankl, Viktor 67
Freud, Sigmund 104, 149, 207, 211
Frith, Chris 240

Galileo 163
Galton, Sir Francis 46, 51
Garbulsky, Gerry 44–7, 50–1, 53–7
Garden of Earthly Delights, The (Bosch painting) 153
Gilbert, Dan 241
Goldberg, Diego 98
Goldenberg, Amit 63–4
Golombek, Diego 69, 163
good conversations
  absence of (bad conversations) *See* bad conversations
  absence of (loneliness) 175, 177–80, 221, 244, 248–9
  advice on 34–6, 70–1, 125
  for conflict resolution *See* conflict resolution
  effects generally 4–5, 11, 35, 175
  good listening 85–6
  good questioning 83–5, 125
  group experiments in *See* group decision-making experiments
  minority views, potentially persuasive force 29–32, 61
  Montaigne on 32, 49–50, 57
  with ourselves *See* compassion
  requirements for 27, 32–3, 35, 41, 49
  in schools 61
  this book's aim for 260–2
Google 114–15, 157
gravity, Newton's law of 19
Green, John 135
Gross, James 206
group decision-making experiments
  on convinced minority view effects 31–3
  on intuitive and approximated deductions 44–7, 50–1
  on Israeli–Palestinian conflict 63–4
  on seven deadly sins ordering 154
  on social pressure effects 28–9, 33
  on taboo subjects and moral dilemmas 53–7
growth mindsets 65–8, 71
"gut feelings" (intuition) 43–4, 96, 132
Gutiérrez del Álamo, Hortensia 193

haka (Maori warrior dance) 186
Halperin, Eran 63–4
happiness 236–7
Harley, Trevor 96, 98
Harris, Sam 251–6
Harrison, George 81–2
Heffernan, Margaret 52
Heidegger, Martin 14
Heine, Steven 140
Henrich, Joseph 140

Henry IV of France 57
Hera, Diego de la 59–61
Heraclitus 79
Hingis, Martina 185
hippocampus 112–13
Homer 82, 117, 182, 204
Honnold, Alex 199, 200
hope 66–7
Hubbard, Edward 136–7
Hulk, the (Marvel Comics character) 181, 189, 204
humor
  benefits of 230, 238–40
  contagion effect of laughter 48, 190–4, 222

identity formation. *See* life story formation
implicit memory 105, 106
improvisation 95
Inception (film) 92
incest debate 53–4
induction
  emotional governance by 184, 186–7, 189–94, 222
  reasoning by 132, 157–9
inert knowledge/learning 89–90, 93–4, 125
infantile amnesia 77, 102–6
Inglourious Basterds (film) 54
Iniesta, Andrés 185, 186
intimidation, effect on conversations 33, 51–2
intuition 43–5, 96, 132
Israeli–Palestinian conflict, group experiment on 63–4
It's All About Love (film) 2
Izard, Véronique 108

James, William 145, 189
Jaspers, Karl 82–3
Jaynes, Julian 82
jealousy 182, 219–20
Jordan, Michael 184, 186
joy 217–18, 238

Kahneman, Daniel 23
Kamitani, Yukiyasu 164
Kant, Immanuel 140
Keller, Helen 278
Kennedy, John F. 41–3, 50
Kennedy, Robert 42
Kober, Hedy 248
Koenig, John 160
Koestler, Arthur 122–3
Köllisch, Anton 117
Kraft, Tara 191, 192
Kuhl, Patricia 138
Kuhn, Franz 144

Landauer, Thomas 156, 157–8, 168
language. *See* words
language acquisition 98–101, 132, 138–9, 156–9
Las Casas, Bartolomé de 33
laughter
  benefits of 230, 238–40
  contagion effect of 48, 190–4, 222
Leary, Mark 245
Ledecky, Katie 184–5, 186
Lee, Kang 16
Lee, Stan 181
Lennon, John 80
Leone, Juliana 133–4
"Let It Be" (Beatles song) 80

Levenson, Robert 238–9
life story formation. *See also* creativity and memory connection
advice on 125–6
editing memories 77–9, 95, 113–14, 116–17, 126
fake news about ourselves 11, 16–20, 77–8
false memories 78, 108–11, 119–24, 126, 165
infantile amnesia and first memories 77, 102–8
self-compassion, effect on 245–6
liget (Maniq emotion, grief at "high voltage") 162–3, 189, 237–8
listening 85–6
literature 13–14
Loftus, Elizabeth F. 111, 120
logic. *See* reasoning
loneliness 175, 177–80, 221, 244, 248–9
love. *See also* compassion
communicating 132–3, 168, 178, 186, 250, 258–9
imprecise meaning of 19, 131, 162
jealousy of lovers 182, 219–20
medieval thinkers on 150, 154
luck 25, 66, 216
lying, psychology of 16

McCartney, Paul 80–1, 83, 88
McDermott, Kathleen 119
Mackay, Charles 47–9
McNeill, David 96
"madness of crowds" 48–9, 51
Majid, Asifa 140–1
Malet, Léo 123–4
Maniq people 140–1, 237
Maradona, Diego Armando 15–16, 18, 110, 111, 120
Marcus Aurelius 34
Mascherano, Javier 184, 186
Matrix, The (film) 89, 236
Mayo, Ruth 192
MDMA (psychotropic drug) 117
meditation 231, 251–6
Mednick, Martha and Sarnoff 120–1
Meltzoff, Andrew 105
memory
and creativity *See* creativity and memory connection
editing memories 77–9, 95, 113–14, 116–17, 126
false memories 78, 108–11, 119–24, 126, 165
from or before birth 106–8, 121
infantile amnesia and first memories 77, 102–8
neurological processes of 111–14, 209
remembering and forgetting 95–8, 106, 114–18, 139
Mercier, Hugo 21–2, 28, 29, 32, 33, 46
Mercury, Freddie 186
Messi, Lionel 160
minority views, potentially persuasive force 29–32, 61
Minos, King of Crete 78
Misanin, James 113
Mithoefer, Michael 117

mnemonics 91–4, 125–6
Mnemosyne 80
Montaigne, Michel de 5, 26, 32, 49–50, 57, 176
"Montaignes"/"confident grays" (consensus makers) 56–7, 63–4, 71
Moser, Edvard and May-Britt 112
Mourinho, José 239–40
movies 14, 212–13
Muses, the 80–2
music. *See* sound
"My Sweet Lord" (George Harrison song) 81–2

Nadal, Rafael 185, 186–7, 192
national stereotypes 20
Navajas, Joaquín (crowd experiments by) 44–7, 50–1, 53–7
Neff, Kristin 241–2, 244–5
negative self-talk 4, 20, 65–6, 188, 227, 230–1, 241
neurology of memory 111–14, 209
Newton's law of gravity 19
Niemann, Christoph 198
Night at the Opera, A (film) 211
9/11 terrorist attacks 115–16
Noah, Tom 192
Noé, Mariana 205
Norenzayan, Ara 140
Nozick, Robert 236–7
Núñez, Carlos 134–5

paradox of choice 86–8, 125
Pashler, Hal 30, 114
Pearson, Helen 25
Peirce, Charles 156
performative statements 19
Perfumo, Roberto 110, 111
persona creation. *See* identity formation
phantom fundamental (auditory illusion) 118–19, 120–1
Piaget, Jean 108–9, 111, 120
pink elephant experiment 209–10
Plato 49, 80, 156, 157, 204
Plutchik, Robert 151–2, 153, 159–60, 167
Polgár, Judit 99
Popper, Karl 17
Posner, Michael 195–6, 207, 211
possessive pronouns 105, 106
post-traumatic stress disorder (PTSD) 116–17
prepositions 99
presence (virtual reality term) 14
Pressman, Sarah 191, 192
problem-solving. *See* reasoning
propranolol (adrenaline-inhibiting drug) 117
Proust, Marcel 139
PTSD (post-traumatic stress disorder) 116–17
Puerto de La Morcuera cycling climb 2–4

Quine, Willard Van Orman 156, 157
Quino 208–9

Ramachandran, Vilayanur 136–7
RAT (Remote Associates Test) 120–1, 157
rational hedonism (Epicurus) 236–7

reappraisal of emotions. *See* emotions, resignification (re-appraisal) of
reasoning. *See also* decision-making
 automatic *versus* logical thinking 11, 20–3, 26, 36
 emotions affecting 31, 36, 208
 fear *versus* reason 68–9, 200–2
 group experiments in *See* group decision-making experiments
 by induction 132, 157–9
 intuition compared 43–4
 lack of evidence undermining 24–6, 109–10, 157
 rational hedonism (Epicurus) 236–7
 social pressure undermining 28–9, 33, 41–2
(re)consolidation processes of memory 111–14, 116–17
reflexivity (self-fulfilling prophecies). *See also* fallibility
 of affectionate gestures 250
 in conflict resolution situations 64–5, 71
 contagion effect of emotions 48, 165–6, 190–4, 222, 230
 of emotion statements 17, 19, 35, 131, 148–9, 168, 203
 fallibility, interaction with 17–19, 26
 growth mindsets contesting 65–8
 of negative self-talk 4, 20, 65–6
remembering and forgetting 95–8, 106, 114–18, 139
Remote Associates Test (RAT) 120–1, 157
repression, emotional governance by 176, 206–7, 209–11, 222
repulsion 215–16
resignification of emotions. *See* emotions, resignification (re-appraisal) of
Richter, Curt 66
right to be forgotten 114–15
Roediger, Henry 119
Romanian orphans 248–9
Romero, Sergio "Chiquito" 184
Rosaldo, Michelle and Renato 162–3, 237
Rosenberg, Erika 194
Ross, Michael 121, 122–3, 124
Roy, Deb 98, 99–100
rubber ducky technique 70
Russell, James A. 150–1, 152, 153
Russell, Jeffrey 58

Sackur, Jérôme 87
sadness
 anger, relationship to 152, 204, 217
 contagion effect of 165–6
 imprecise meaning of 131, 160
 purpose and resignification of 216–17, 218, 238
Sagan, Carl 121
Saint-Exupéry, Antoine de 135
Sanchez Vives, Mavi 14

Sardon, Mariano 133–4
schizophrenia 82
Schlesinger, Arthur 42
Sehgal, Parul 182
self-compassion. *See* compassion
self-fulfilling prophecies. *See* reflexivity
self-help advice 34–6, 70–1, 125–6, 169–70, 221–2, 257–9
semantic memory 105
Seneca the Younger 261
sentences, judicial 19
September 11 terrorist attacks 115–16
Sepúlveda, Juan Ginés de 33
seven deadly sins 153–4, 197
Shaw, Sebastian (Marvel Comics character) 187, 188
Shulgin, Alexander 117
sibling incest debate 53–4
Sigman, Mariano
  childhood move to Spain 101–3, 249–50
  experimental studies 44–7, 50–1, 53–7, 59–61, 133–4
  fear of falling coconuts 200–1
  fictional friendship with Maradona 15–16, 18
  losses of perspective remedied with affection 247, 249–50
  musical achievements 124, 232–3
  sporting achievements 1–4, 124
  televised jump from bridge 69
  uncertain Jewish identity 61–3
Simner, Julia 96
Simonides of Ceos 91
smells, words describing 139–41
social media
  bad conversations on 33, 35–6, 48–9
  distraction caused by 183, 187, 197
  erasure of postings on 114–15
  fake news on 12–13
  "like" economy 86
social pressure, effect on reasoning 28–9, 33, 41–2
Socrates 34
Soros, George 17–18
sound
  phantom fundamental (auditory illusion) 118–19, 120–1
  remembered melodies 80–2
  shape of 135–7
  of words 136–7, 138–9, 141–2, 148, 152
space and time perceptions 133–5, 137, 148
Spanos, Nicholas 121
Spelke, Liz 108
Spider-Man (Marvel Comics character) 20
Spiegelman, Art 66–7
sport, mental dimension 1–4, 67–8, 150, 184–7, 188–9, 217, 234
Squire, Larry 105
Stepper, Sabine 190–1
stereotypes 20
stories about ourselves. *See* life story formation
Strack, Fritz 190–2, 194
Strange, Dr. Stephen (Marvel Comics character) 44

stress
  distraction and repression causing 176, 197–8, 206–7, 222
  lack of affection causing 249
  loss of perspective causing 233–4, 247
  and memory 116–18
  methods of relieving 87, 191, 192, 222, 244–5
success 25, 217, 234, 235
Sullivan, Anne 278
suppression, emotional governance by 176, 206–7, 209–11, 222
surprise 216
symposia 5, 49

Taminiau, Jan 63
theory of mind system 212–13
Theseus's ship 78–9, 102, 121–2
Thomson, J. K. 206
time and space perceptions 133–5, 137, 148
tip-of-the-tongue phenomenon 96–8, 106, 114
Todorov, Tzvetan 33
Tonegawa, Susumu 114
Treisman, Anne 109, 110, 118
Tulving, Endel 105
12 Angry Men (film) 29–31, 32, 33, 61
Twitter (now X) 12–13, 49

Ulysses pact 182, 183, 187

Valladolid debate (1550) 33
vertigo 199–200
virtual reality 13–14
Vosniadou, Stella 59, 61
Vosoughi, Soroush 12–13
vowel sounds 136–7, 138–9, 141–2, 148
Vul, Ed 30

Ward, Jamie 96
Watts, G. F. 142
Whitman, Walt 30
Wilkins, John 142–5
Williams, Serena 147, 152, 185, 186
Wilson, Anne 121
"wisdom of crowds" 44–6, 47, 51
Wittgenstein, Ludwig 156
words
  definitions, geometry of 155–9
  describing emotions *See* emotions, words describing
  describing smells 139–41
  learning 98–101, 132, 138–9, 156–9
  memory formation role 104, 105–6
  remembering 95–8
  sounds of 136–7, 138–9, 141–2, 148, 152
  time and space expressed in 133–5, 137, 148
  word association experiments 119, 120–1, 136–7, 157
"World Is Wide and Strange, The" (Borges and Bioy Casares short story) 132
writing
  and conversation 47, 49, 70
  of memories 82–3, 91
  sheet music 135–6

X (formerly Twitter) 12–13, 49
X-Men (Marvel Comics characters) 187, 255

"Yesterday" (Beatles song) 80–1, 83

Zermelo, Ernst 87–8